Anonymous

Curtis' Bay

Its superior advantages and admirable location as the only existing and

available deep water harbor contiguous to the city of Baltimore

Anonymous

Curtis' Bay
Its superior advantages and admirable location as the only existing and available deep water harbor contiguous to the city of Baltimore

ISBN/EAN: 9783337064471

Printed in Europe, USA, Canada, Australia, Japan

Cover: Foto ©berggeist007 / pixelio.de

More available books at **www.hansebooks.com**

CURTIS' BAY;

Its Superior Advantages and Admirable Location

AS THE

ONLY EXISTING AND AVAILABLE

DEEP WATER HARBOR

Contiguous to the City of Baltimore,

IN CONNECTION WITH ITS

RAPIDLY INCREASING LOCAL MANUFACTURES, THE
DEVELOPMENT OF ITS COAL TRAFFIC, AND
THE ACCOMMODATION OF ITS

Western and Southern Railroad Connections.

PUBLISHED UNDER THE AUSPICES OF

THE PATAPSCO LAND COMPANY

OF

BALTIMORE CITY.

PRINTED BY JOHN MURPHY & CO.

182 BALTIMORE STREET, BALTIMORE.

List of Directors.

JOSEPH W. JENKINS,   *Baltimore, Md.*
JOSHUA HARTSHORNE,   "   "
WILLIAM S. RAYNER,   "   "
HIRAM KAUFMAN,   "   "
WILLIAM C. PENNINGTON, "   "

General Officers.

WM. C. PENNINGTON, *President.*

JOSIAS PENNINGTON, *Sec'y and Treas'r, Pro tem.*

OFFICE, No. 4 LEXINGTON BUILDING, S. W. CORNER OF CHARLES AND LEXINGTON STREETS.

INTRODUCTION.

NEARLY half a century has elapsed since the attention of the merchants of Baltimore was fully directed to the necessity for providing easy and expeditious lines of communication between the city and port, with which they were identified and the then infant West, if they wanted to perpetuate the commercial importance of Baltimore, and neutralise the prejudicial influence to their interests exercised by the public works in Pennsylvania and the Erie canal in New York. They realized that a large amount of business which was legitimately tributary to their city, was being gradually attracted to New York and Philadelphia, in consequence of the superior facilities for transportation furnished by the States of which these two cities were the recognized commercial centres, and that their bright anticipations for the future of the "Monumental City" would be comparatively blasted unless they could compete on more equal terms with their rivals, and furnish for their customers some more expeditious and certain means of transportation than those which were coincident to the old fashioned wagon and stage routes. At the time alluded to, a project for building the *Chesapeake and Ohio Canal* had been inaugurated and work had been commenced, but its success as a commercial undertaking was seriously questioned in consequence of the high elevations over which it had to be carried, and the scarcity of water, while its projected eastern terminus at Georgetown on the banks of the Potomac was calculated to prejudice in many respects the commercial interests of Baltimore, or at any rate exclude its citizens from the actual benefits which had been anticipated from the construction of the canal. This fact was so fully realized by a then prominent merchant of Baltimore, Mr. Philip E. Thomas, that he voluntarily withdrew in 1827 from the canal commissionership which he held as representative for the State of Maryland, and applied himself

energetically in connection with Mr. George Brown and other
prominent citizens of Baltimore, to maturing a plan for building a
railroad which should be the highway for traffic from the Ohio
River at Wheeling to Baltimore. The ideas as then enunciated
by Messrs. Thomas, Brown and their associates, appeared at the time
somewhat chimerical because although short lines of railroad, such
as the "Stockton and Darlington" in England and the "Granite
Branch" near Boston, had been built for the conveyance of coal
and stone to navigable waters, no railroad had been constructed
either in Europe or in this country for the general conveyance of
passengers and produce between distant points :—In brief for general purposes, railroads were regarded as an untried experiment, and
it was a mooted question whether horses or stationary steam engines
would be the preferable motor.   In laying their plans fully before
the citizens of Baltimore at that time, and corroborating their views
by the opinion of prominent engineers both in Europe and America,
the committee alluded at length to the advantages possessed by the
city of Baltimore as being 200 miles nearer to the navigable waters
of the West than New York and 100 miles nearer than Philadelphia, also that the easiest and by far the most practicable route
through the ridges of mountains which separated the Atlantic from
the western waters was along the depression formed by the Potomac
River in its passage through them.   Special allusion is made to
these historical facts, as indicating the prescience and business
sagacity of the Baltimore merchants, and although it is not within
the sphere of our present duties to trace out the various steps by
which the Baltimore and Ohio Railroad Company, (incorporated
in accordance with the views of Mr. Thomas and his associates in
1828,) successfully overcame every obstacle which nature, the hatred
of innovation and political chicanery placed in their path; still the
inhabitants of the city of Baltimore and the State of Maryland
may be proud of having been the pioneers of railroad construction
in this country, of having been practically the founders of a system
which now extends in an unbroken line from the Atlantic to the
Pacific; which has pierced mountains and spanned the mighty
rivers of this vast continent, and which, whether regarded as the
missionary of civilization or the architect of industrial development and material prosperity has tended to make the United States

one of the foremost among the nations of the world, to be respected
and honored by the dynasties of either hemisphere. These wise
and good men, who projected a great and glorious future for the
city of Baltimore, and for the State of which it is the recognized
commercial centre and manufacturing emporium, rest in their
honored graves, but the memory of their noble aspirations lives
fresh and green as the flowers which deck their last homes in the
hearts of the present generation, whose aim seems to be that Baltimore shall avail itself by every legitimate means of the superior
geographical advantages which it possesses, and that to its port
shall converge not merely the traffic originating on the waters
west of the Alleghanies, but the trade of the Western, South-western and North-western States and Territories, the older trans-
Atlantic countries and of the West Indies and South America.

It is true that the present proud position of Baltimore in connection with the commerce of America, is mainly due to the con-
servative and consistent manner in which successive managers of
the Baltimore and Ohio Railroad Company have adhered to the
policy of their predecessors; and by advancing, *pari passu,* with
the growth of the city in the development of new railroad enterprises, have secured for the railroad proper, as well as for the
mercantile community, large accretions of wealth; but a question
naturally arises at this juncture, when the cap stone, (to speak
figuratively,) is shortly about to be placed on the scheme and system of the Baltimore and Ohio Railroad Company, by the completion of its new line from Centreton on the Lake Erie Division,
to Chicago; and when other railroad lines such as the Baltimore
and Potomac, Northern Central and Western Maryland are re-
quiring terminal accommodations at tide-water for a rapidly increasing coal trade and general traffic, whether the necessary
facilities for handling such a large accretion of traffic, as will be
coincident to the growth of the West and the proper development
of local business, can be furnished either at Locust Point, the
present tide-water terminus of the Baltimore and Ohio Railroad,
or at Canton where it was anticipated, (although such anticipations
have, in consequence of prohibitory rates, not yet been realized,)
that the tide-water business of the Northern Central and Western
Maryland Railroads would be concentrated. It is the object of

this present pamphlet to show, and it is hoped conclusively, that
other provisions must be speedily made for accommodating the
prospective traffic of Baltimore and for developing its various
industries. Should the population and commerce of the city increase in the same ratio during the present, as in the past decade,
or should an additional stimulus be given to the manufacturing
industries, as is now indicated by an evident appreciation of the
geographical advantages of Baltimore in all sections of the country, the existing terminal facilities of the Baltimore and Ohio
Railroad Company would not be adequate to their requirements,
and they would be compelled to seek additional outlets to deep
water contiguous to their main line of road where the handling of
produce, general merchandise and coal traffic could be conducted
economically and expeditiously. The various reasons why a location at Curtis' Bay on the property of THE PATAPSCO LAND
COMPANY would be advisable for the railroad, commercial and
manufacturing interests are herewith given ; and, in connection
with the map appended to the pamphlet should convince the capitalists of Baltimore, that the elaborate scheme of improvements
now contemplated by the company owning the property is destined
to attract a large increase of capital to Baltimore, to build up its
commercial supremacy, and enable it to compete on more than
equal terms with the rival cities of Boston, New York and Philadelphia ; in fact, the new town of *Pennington,* to be erected on
the Curtis' Bay property, is in all probability destined to become
for Baltimore what Brooklyn, Jersey City and Hoboken have been
to New York City.

That these facts have been fully appreciated by some of the
leading business men of the city, is evidenced by the fact that
the property now about to be improved and adapted for a port,
has been held by its present owners for nearly a quarter of a
century ; it has been held for such a term of years with a firm
confidence that its merits as a shipping point would be ultimately
appreciated, and that, by retaining its control, they would be instrumental in developing the geographical advantages of which
nature had made them the possessors, and in which their fellow-
citizens were so deeply interested, as enabling them to carry out
well conceived plans for the commercial preëminence of Baltimore.

To render the comprehension of the value of this property perfectly intelligible, it is proposed to allude in detail to the following subjects:

(1.) BALTIMORE, its early foundation, growth and present condition.

   (2.) Advantages of location for commercial and manufacturing purposes.

   (3.) Situation of harbor.

   (4.) Railroad connections.

   (5.) Present terminal facilities.

   (6.) Difficulties in the way of further development at existing termini.

   (7.) Curtis' Bay: its advantages as a port and as a manufacturing centre.

   (8.) Present plan of THE PATAPSCO LAND COMPANY OF BALTIMORE CITY, for developing their property.

(I.) BALTIMORE—ITS EARLY FOUNDATION, GROWTH AND PRESENT CONDITION.

THE town of Baltimore, with sixty acres of ground, was founded in 1729, on the north side of the Patapsco, and was named after Lord Baltimore, to whom the State of Maryland was originally granted in 1633. The country had, however, been previously discovered by a Captain John Smith, in 1606, and there was a population in Maryland, about 1659, amounting according to the most reliable returns, to more than 12,000. Tobacco was the great product of the province at that early date, and the ancient chronicles assert that a hundred sail of ships a year from the West Indies and from England, traded in this article, which to use the phraseology of that period, was the source of a very large revenue to the English crown, at a vast expense, industry and hazard to the Lord of Baltimore. In 1706, an act was passed creating "Whetstone," now better known as Locust Point, a town; but no definite progress towards definitely locating the town appears to have been made under the provisions of this act, and in 1729, efforts were made to found a city on the property which looks in on Spring Gardens, then owned by a Mr. John Moale, a merchant from Devonshire, but the project was strenuously opposed by him through the belief that his property was rich in iron ore, and he had sufficient influence with the Legislature, of which he was a member, to defeat the plan. From these causes the original founders of Baltimore were compelled, against their own wishes, to abandon the level land, and seek a location for their embryo city, under the hills and amid the marshes of the north-western branch, where Charles and Daniel Carroll had agreed to sell certain property amounting to about sixty acres. In 1745, Jones Town was incorporated with the city of Baltimore, under the latter's corporate name, and in 1747, a then unoccupied portion of land, about eighteen acres, lying between Baltimore and Jones Town was absorbed into the municipality. Twenty-two years after

Baltimore had been incorporated, viz: in 1751, the commercial requirements of the city had increased so rapidly that the erection of a market house and town hall was deemed advisable. The war between the English and French in the few succeeding years had a tendency to increase the population of the town of Baltimore proper by inducing the inhabitants of the State to remain in the older settlements, and not penetrate into the sparsely settled interior where they were subject to annoyances and attacks from the hostile forces, and more especially from the Indians. In 1776, the population of Baltimore received considerable accessions from the refugees from Nova Scotia, and there was a perceptible growth in the city developments and in its manufacturing industries. Ship yards were established at what was then known as Fell's Point, and a large trade for the then infant province was carried on more especially in tobacco, but in lesser proportion in wheat, lumber, corn, flour, pig and bar iron, skins and furs. The exports of tobacco alone from Maryland to England, were estimated in 1763, to be about 28,000 hogsheads annually, valued at £140,000, and the greater proportion of this trade paid tribute directly and indirectly to the mercantile enterprise of Baltimore. Nothing indicates more clearly the general prosperity of the province and its rapid development in thirty years after the foundation of Baltimore, than the increase of population during that period. In 1733, (according to returns given by Mr. George E. Howard,) the taxable population, (including males above the age of sixteen, and all negro and mulatto females,) numbered 31,470. Fifteen years afterwards the entire population was 130,000, (94,000 whites and 36,000 blacks.) In 1756, it had increased to 154,188, (107,963 whites and 46,225 blacks.) In 1761, it amounted to 161,307, (114,332 whites and 46,975 blacks.) It is not within the province of this pamphlet to trace the various steps by which the ill-advised home government attempted by restrictive legislation and prohibitory measures, to stifle the growth of manufactures in the province of Maryland, and thereby render one of her colonies dependent on England for all its trade in manufactured articles; nor to show how the formation of a mercantile marine was stopped by restricting the trade of the province to English ports, and insisting that such trade should be only carried in English bottoms.   It is irrelevant also to show how this protective and prohibitory system crushed out any spirit of servile adulation and compromise which might have resulted from a temperate recognition of just claims by the then British administration. These and kindred topics have been ably and successfully handled by historians of the past and present, suffice it to say that the very action which was taken to repress independence only added fuel to the fire, it developed in the early settlers of this country an energy of character and a spirit of honest reliance and manly virtue, which handed down to posterity and cherished as valuable heir looms, bid fair unless the temper and disposition of succeeding generations are warped by an excess of prosperity to make the United States in social and mental characteristics as it is in the publicly and varied character of its resources, the first country in the world.   An active and independent population thrown on its own resources soon adapts itself to the new situation of affairs; hence, it is not surprising to find that in 1778, factories and mills for the manufacture of linen, woolen goods, nails, paper and iron were at work in Baltimore; fast sailing traders were built and a considerable traffic was carried on by them during the Revolutionary war with the West Indies; in fact, even under what would have appeared to most people, very unfavorable auspices, the commerce of the town of Baltimore increased so rapidly that in 1780 a custom house was established, and merchants were relieved from the annoyance and inconvenience attendant on entering and clearing their vessels at the port of Annapolis as had been formerly customary.

A fresh stimulus was given to the commercial enterprise and activity of Baltimore by the French revolution and the protracted war resulting therefrom, which devastated the continent of Europe, and by the interruption of agricultural pursuits, caused a greater demand for American wheat and flour. The colonial dependencies also of the various conflicting powers were forced, by being cut off from their home connections, to open up trade with a neutral, and America profited immensely by becoming a market for the sale of produce, and for the purchase of necessary supplies. It was at this time, more especially from 1790 to 1801, and again from 1803 to 1812, that Baltimore became the recognized entrepot.

for traffic between the West Indies and all ports of Europe, and, notwithstanding the heavy risk incurred in blockade running, vessels built on the Chesapeake were uniformly successful in evading the cruisers of every blockading squadron, and in transacting a remunerative business. Nor were the growth and prosperity of Baltimore fostered merely by a foreign carrying trade—emigration had commenced to flow slowly but steadily towards the West, and it was found that the geography of the country plainly indicated Baltimore to be the original and natural terminus of internal American trade on the Atlantic seaboard. Hence we find that, even at that early date, a considerable traffic was carried on with the embryo settlements on the navigable waters of the West. It is true that this traffic was carried on at, what will seem to us at the present day, considerable disadvantage, and there was great delay in transportation, but the pack horses of the revolutionary period were quickly superseded by the cumbrous six or eight horse covered wagon, the narrow and circuitous paths along which the pack trains moved along in single file were supplanted by the substantial turnpike, and the old "Braddock's Road" will be in succeeding generations as suggestive a memento of the commercial enterprise of Maryland and her merchants as the canals and railroads. It may be appropriately noted here, that in 1796—sixty-seven years after its original foundation—the town of Baltimore was elevated to the dignity of a city, and a charter of incorporation, under the name of the "Mayor and City Council of Baltimore," was granted by the State Legislature. The statistics of the United States census at this time furnish data relative to the growth of the city, which may be interesting. In 1790 the population was 13,503; in 1800, 31,514, and in 1810, 46,555, an increase of nearly 350 per cent. A temporary check was given to the development of the city of Baltimore, by the war with England, from 1812 to 1814; also, by the establishment of peace throughout Europe in 1815 and a withdrawal to the ships of the various trans-Atlantic nationalities of the traffic which had been carried for some time in American bottoms; still, there was a continued increasing demand, during several years, for American wheat and other produce, while new traffic with South America, and more especially with the newly established empire of Brazil, compensated for any diversions of business into other channels. It could not be expected that Baltimore should have been exempted from the disastrous financial complications resulting from the establishment of the United States Bank,—complications which culminated in 1819, and which involved many enterprising mercantile firms and individual subscribers in utter ruin and penury. It was also disastrously affected by the panics of 1837 and 1857, and its commerce, which had thriven immensely under exceptional causes, from 1790 to 1815, did not show a corresponding increase in the forty-five years immediately subsequent to the pacification of Europe; but still, there was a steady and permanent growth, and the population which, in 1820, was 62,738, had risen, in 1860, to 212,418, as will be seen from the following returns of the United States census:

Year.	State of Maryland.	County of Baltimore.	City of Baltimore.
1820	298,260	96,201	62,738
1830	399,455	120,870	80,625
1840	501,793	134,379	102,313
1850	583,034	210,646	169,054
1860	687,049	266,553	212,418

It might be well, before proceeding further in the history of the growth of Baltimore, to note that in the forty-five years ending 1860, alluded to above, her name is prominently associated with three of the most important improvements of the present century. Baltimore was the *first* city in America, according to the most reliable records, which was lit with gas—viz: in 1816. The citizens of Baltimore were, in 1827, the *first* to inaugurate a system of railroads for the transportation of passengers and general merchandise, while, seventeen years later, in 1844, it was between Baltimore and Washington that the *first* electric telegraph, not merely in America, but in the *world*, was erected. Had the city, during that period of nearly half a century, inscribed nothing else on her chart of progress and material development than these three isolated facts, she would be justly entitled to the gratitude of civilized America and the panegyrics of millions.

A severe, although temporary check to the growth of Baltimore resulted from the civil war, which lasted from 1861 to 1865. Traffic with the South was entirely suspended; industries on which the inhabitants had been entirely dependent were paralyzed, and commercial relations with the West were interrupted and par-. tially diverted to other cities. It is not our province to criticise the causes which brought about this civil war, nor the attitude which was assumed by the majority of the inhabitants of Maryland at the time of a most eventful crisis in the nation's history; suffice it to say, that great allowance should in every case be made for early education, sympathy and associations. If the people of Baltimore erred in their understanding and interpretation of that great contest; if they leaned more to the Confederate than the Federal cause, it must be candidly admitted that such errors (if they were errors) were conscientious and in accordance with their interpretation of right and duty—the convictions on which their actions were based were honest—the proclivities which influenced them were deep-seated and meritorious ; and the records of both contending armies show how many noble spirits sealed their faith, to write figuratively, with their life-blood,—how the fervor and manly courage of their ancestors, as recorded on many a battle-field during the Revolutionary war, had been indelibly impressed on the scions of a succeeding generation. There is no doubt that Baltimore suffered very severely from the frontier position which was occupied by the State of Maryland, and that although her soil was on but two occasions the field of battle between the armies of the North and South, still a general tone of depression and demoralization was engendered by the continued presence of troops within her territories, and by the continued suspicion of sympathy with the South to which the citizens were unfortunately subjected; but during this period of forced commercial inactivity, except in so far as Government contracts were concerned, plans were being matured for developing the trade and commerce of the city; new avenues of traffic were sketched out ;—the business of the West might have been temporarily diverted, but it had not been irrevocably lost ;—the South might be temporarily paralyzed, but its recuperative energy was still on a par with that of other sections of the American continent, and until it regained in happier times its normal condition of prosperity and affluence there was a large possible European commerce to be built up ;—the local resources of the city and State were to be developed, and a fresh impetus might

be given, by a correct representation of geographical advantages,
to the investment of foreign capital and to the immigration of
skilled labor.   The plans thus matured have not proved chimerical ;—already the tide is turning in favor of Baltimore, the grana-
ries of the West are pouring their riches along the well-developed
arteries of transportation into the elevators erected by the Baltimore and Ohio Railroad Company at Locust Point; other products of the Mississippi Valley and the great North-west respond
to the facilities furnished and converge to Baltimore, while the
fleet of German and English steamers constantly plying between
this and European ports testifies that the ideas which induced the
founders of Baltimore town to prophesy its commercial supremacy
were by no means exaggerated, and that the natural geographical
laws of location remain *immutable*.

The following statistics indicate very clearly the truth of these
statements :

	1860.	1870.	1871.	1872.
Imports at the Port of Baltimore	$9,784,773	$21,017,313	$26,770,181	$29,429,439
Exports of Domestic Merchandise	8,084,606	12,396,518	18,236,166	17,381,591

The census returns for 1870 show that the industrial products
of the city and county of Baltimore amounted to the sum of
$59,219,993, in which was employed a capital of $26,040,040.
The population of the city proper had increased from 212,418 in
1860, to 267,569 in 1870, and a school census, taken in October,
1873, proves that it has now risen to 319,000.   The assessed
value of property at the time of taking the census in 1870 was
$237,806,530, and its real value $401,634,738; the tonnage of the
port was at the same time 150,086, and the city debt $13,568,431,
or at the rate of about $51 per capita of population.

And yet Baltimore is in its comparative infancy, the commercial growth of the past eight years merely indicates its capabilities
for further development when its superior advantages of economical transportation are more thoroughly known and appreciated by
the producers of the West and by the consumers of the East; when
it becomes, as in former years, the trading mart of the South, and
when all sectional differences have been obliterated by the interchange of progressive ideas and by a growing similarity of interest.

More weight will perhaps be attached to these brief remarks on the growth and present condition of Baltimore, when it is known by the reader that they emanate from one who is not identified with the city, but who has been compelled, in the course of his professional duties and literary avocations, to study carefully, and it is hoped impartially, the relative merits of different cities in the Union, now engaged in a healthy competition for commercial supremacy and prestige. Those who are blinded by the prejudices of early education and associations with certain localities, may fail to recognize the geographical fact that Baltimore is NEAREST the North, NEAREST, the South, NEAREST the West, in fact, so central on the seaboard as to be NEAREST all classes of industry and of production. They may attempt to ignore the fact that it is NEAREST the manufacturer of the North, the producer of the West, the cotton planter of the South and the purchasers of Europe and the West Indies or South America; they may claim that the capital or influence of other States and cities can divert traffic from its ordinary geographical short lines of transportation into more circuitous routes, but any *correct* and *truthful* not *distorted* map will show them that Baltimore is the *natural*, not artificial, depot of internal traffic, and that the trunk line, with which its history is so closely identified, is, at any rate for the present, the SHORTEST and MOST DIRECT avenue of communication between the West and the Atlantic seaboard, thence to Europe. The merchants and public men of Baltimore will be strangely recreant to the principles which animated their forefathers, if they fail to avail themselves of their present vantage ground, and do not anticipate the commercial requirements of their city. The solid men should come to the front. The press, of whatever shade of politics, should be a unit in laboring for the development of Baltimore and its commercial and manufacturing importance, and provision should be made at once to accommodate an import and export trade amounting, within the next five years, to more than eighty million dollars, annually, in the aggregate.

Allusion has now been made to BALTIMORE, its early foundation, growth and present condition; it has been shown, and it is hoped satisfactorily, that few, if any, cities of the Union can show a similar percentage of increase in population, manufactures and

2

commercial and industrial resources as Baltimore: but it would be wrong to leave this topic without mentioning briefly other than geographical advantages which Baltimore possesses for attracting population and consequently wealth. Prominent among these advantages may be mentioned the *climate*, which, as the city is situated about the centre of the Atlantic coast, is not subject to the intense cold of Northern latitudes, nor to the tropical eccentricities of the more Southern States; in brief, it is equable, balmy and healthy, while the location of the city, on a succession of hills rising gradually from the harbor, prevents in a very marked manner the occurrence of those mephitic and miasmatic vapors to which other cities not so advantageously located are subject. In addition, Baltimore is naturally well drained and sewered; a heavy rain must carry off all the impurities of the streets, and in the absence of such rain there is enough water available for city purposes to compensate for any irregularity of nature. Again, apart from the climate, which is all that could be desired by even the most fastidious, and apart from the easy and moderate rentals which must tend to build up the population of Baltimore, there is its Public School system which in providing the best means of education, strengthens the moral influence of the city and educates the rising generation, not merely in the habits of the past, but in those ideas of the future which should be coëxistent with its growth to man's estate.

(II.) BALTIMORE—ITS ADVANTAGES OF LOCATION
FOR COMMERCIAL AND MANUFACTURING PUR-
POSES.

Prior to the outbreak of the rebellion, the commerce of Baltimore was much less developed than now. It is true that even
then the Baltimore clippers had a world-wide renown, and were
constantly employed in the trade to Havana and other South
American ports. It is true that in the sale and manufacture of
tobacco, as well as in the packing of oysters, fruits, vegetables,
&c., the "*Monumental City*" had acquired an enviable reputation. A large trade also was transacted with the merchants
of the South, and indirectly with the section of country lying
West of the Alleghanies; but no systematic efforts were made to
render Baltimore a commercial metropolis and a centre to which
the traffic of all the Southern cities and the West, as well as
Europe, should radiate, until 1866, when the Baltimore and Ohio
Railroad Company determined to complete the policy consistently
pursued by its former managers, and by developing its Western
connections; also, by the erection of elevators and the establishment of a steamship line between Baltimore and Europe, inaugurate a new era in the history of the commercial metropolis of Maryland. All these movements, however, would have been unavailing
had not nature furnished, in the geographical location of the city,
advantages which could not be ignored for commercial and manufacturing purposes. Situate near the Chesapeake bay, on the
Patapsco river, with a climate unsurpassed for salubrity, with a
hygienic record superior to that of any other large city in the
United States, with abundant capital available for the development of all legitimate enterprises, and with a population equally
allied in sentiment and consanguinity with the Northern and
Southern States, Baltimore could not be backward in progress and
material development;—she was forced to recognize the claims
which nature made on the enterprise and perseverance of her citi-

zens, and to labor in the creation and perpetuation of a brilliant and prosperous future. The truth of these remarks will be readily recognized by reference to the map of the United States, where it will be seen that Baltimore is the most accessible port for Petersburg, Norfolk, Richmond, Wilmington, N. C., Charleston, Savannah, Key West, Havana, New Orleans and Galveston. Regular lines of steamers are now in successful operation to the ports above mentioned, while a large traffic is carried on by canal and ocean to Philadelphia, New York, Providence and Boston. It is not, however, merely by her water facilities that Baltimore can claim the palm for superior geographical location. Had she been dependent on these alone, without possessing short-rail line advantages, commercial preëminence would have been in every respect a failure; in fact, the economies of distance, from the recognized centres of trade in favor of Baltimore, as compared with New York, are so thoroughly recognized by the Federal Government that $300,000 was appropriated by Congress in 1871 and 1872, with great unanimity, for deepening the channel, and $65,000 for supplying range lights of the most improved description for the approaches to the harbor of Baltimore. The comparative economics of distances alluded to will be apparent from the following table:

COMPARATIVE DISTANCES TO BALTIMORE, NEW YORK AND PHILADELPHIA.

FROM PITTSBURG, PA.,

To Baltimore via Baltimore and Ohio Railroad	327 miles.
" New York via Pennsylvania Railroad	431 "
" Philadelphia via Pennsylvania Railroad	351 "

Difference in favor of Baltimore as against New York	104 miles.
" " " , " " Philadelphia	27 "

FROM CINCINNATI, OHIO,

To Baltimore, via Baltimore and Ohio Railroad	589 miles.
" New York, { via New York Central Railroad	882 "
via Erie Railway	801 "
via Pennsylvania	744 "
" Philadelphia, via Pennsylvania Railroad	667 "

Difference in favor of Baltimore as against average distance to New York	240 miles.
" " " " " distance to Philadelphia	78 "

FROM LOUISVILLE, KY.,

To Baltimore, via Baltimore and Ohio Railroad	696 miles.
" New York, { via New York Central Railroad	989 "
via Erie Railway	987 "
via Pennsylvania Railroad	851 "
" Philadelphia, via Pennsylvania Railroad	774 "

Difference in favor of Baltimore as against average distance to New York...	246 miles.
" " " " " distance to Philadelphia	78 "

FROM CHICAGO, ILL.,

To Baltimore, via Baltimore and Ohio Railroad	795 miles.
" New York, { via New York Central Railroad	980 "
via Erie Railway	961 "
via Pennsylvania Railroad	899 "
" Philadelphia, via Pennsylvania Railroad	833 "

Difference in favor of Baltimore as against average distance to New York...	132 miles.
" " " " " distance to Philadelphia	18 "

FROM ST. LOUIS, MO.,

To Baltimore, via Baltimore and Ohio Railroad	929 miles.
" New York, { via New York Central Railroad	1167 "
via Erie Railway	1201 "
via Pennsylvania Railroad	1050 "
" Philadelphia, via Pennsylvania Railroad	973 "

Difference in favor of Baltimore as against average distance to New York...	210 miles.
" " " " " distance to Philadelphia	44 "

From all points south of Baltimore the distance in favor of Baltimore is 200 miles. With these short line advantages in its favor over one of the recognized grand Trunk lines of the country, with the lumber, coal and general traffic of Pennsylvania and Western New York converging to it over the line of the Northern Central Railway, with the almost inexhaustible coal supplies of the Cumberland Basin tributary to the Western Maryland Railroad, brought even now to the city limits, and with the large prospective business of the South, transported over the Baltimore and Potomac and Washington Branch Railroads, the city of Baltimore has, undeniably, advantages of location for commercial progress and development; advantages, which can only be limited by the ability of its capitalists and business-men to appreciate fully the situation and further the march of progressive improvement by temperate and well advised investments.

It may be appropriately noted in this connection, that during the ten years ending September 30th, 1873, the Through Tonnage of the Baltimore and Ohio Railroad increased from 166,118 tons, to 640,265 tons, or very nearly 400 per cent.; also, that during the four years ending September, 30th, 1873, the *through* as distinct from *local* tonnage showed an increase of more than 220 per cent., viz: from 286,835 tons in year ending September 30th, 1870, to 640,265 tons in fiscal year ending September 30th, 1873. The following statement is also indicative of the business growth of the city, and its rapid commercial development:

Baltimore and Ohio Railroad.	1872.	1873.	*Increase.*
Through Merchandise East and West..................Tons, 557,609.............		640,265.............	82,656
Corn and other Cereals...............................Bushels, 6,049,430.............		7,510,657.............	1,461,227
Barrels of Flour..757,842.............		940,027.............	182,785
Live Stock...Tons, 72,631.............		87,660.............	15,029

The coal traffic increased during the same period 358,459 tons, amounting to 2,019,718 tons in fiscal year ending September 30th, 1873, as against 1,661,259 tons in 1872; and, while from exceptional causes there was a decrease in the lumber traffic for 1873 of 5,161 tons as compared with the previous year, still the tonnage from that source was 9,292 tons in excess of year ending September 30th, 1871.

Attention might also be drawn to the fact, that the only existing drawback to the commerce of Baltimore, viz: the danger of navigation round Cape Charles will shortly be obviated by the construction of the " Maryland and Delaware Ship Canal," across the peninsula of the states of Maryland and Delaware uniting the waters of the Delaware and Chesapeake Bays. The company which proposes to construct this canal was chartered by the General Assembly of the State of Maryland in April, 1872, and by the Legislature of the State of Delaware in March, 1873. The capital stock authorized by the Maryland charter is $2,000,000, with power to increase to $4,000,000 in case the authorized issue of $4,000,000 six per cent. bonds does not prove sufficient to complete the canal. One route has been surveyed for this new canal called the "Sassafras Route," and the distance from the point where the navigation of the Sassafras river ends, to the mouth of Blackbird Creek on Delaware Bay is seventeen miles. The headwaters of the Sassafras river and Blackbird creek meet in the centre of the peninsula, and formerly the tides flowed within three miles of each other and it seems the natural course for a ship

canal; but there is some diversity of opinion relative to the respective merits of the Sassafras and Chester rivers in connection
with an outlet to Chesapeake Bay, and pending that decision work
has not been yet commenced. The advantage of this ship canal
as now projected cannot be over estimated, in view of the fact that
it lessens the distance by water communication from Baltimore to
all the northern and eastern and European ports about 225 miles;
in addition to avoiding the dangerous navigation around Cape
Charles, alluded to above. Commerce will naturally be attracted
to the point where capital exists and where there is cheap transportation. Of the presence of capital in Baltimore, and of its
more equal distribution among the community there than in any
mercantile city of the same class there can be no doubt, and if, by
the construction of this ship canal, coal, iron, lumber, lime, tobacco and flour, which constitute extensive shipments can be
shipped to New York at ninety cents less per ton than by the old
Chesapeake and Delaware canal, which is circuitous in its route,
is worked by horses and is subject to detention by locks, the
commerce of Baltimore will be considerably aggrandized. Allusion has been made to the fact that commerce will naturally be
attracted to the point where capital exists and where there is cheap
transportation; justice to the public-spirited individuals who have
fully appreciated the correct geographical position of Baltimore as
a commercial centre, demands that we should notice in this connection some elements of superiority over New York which cannot be
ignored; and in mentioning New York, we do so on the ground
that it has been erroneously assumed to be the grand objective
point for all foreign business. In New York, steamers arriving
from Europe are subjected to heavy port charges, in addition to a
large annual rental for wharves; freight destined for inland points
has to be carted from the steamship pier to the receiving depot of
the railroad on which the consignee is located; otherwise, if there is
any delay in passing it through the custom house, it is mercilessly
consigned to some bonded warehouse, and after protracted delay,
with numerous charges affixed, reaches its destination. Again,
emigrants arriving at New York are sent to Castle Garden from Hoboken, Jersey City or whatever pier the European steamer is docked
at; they are exposed for some days in the majority of instances to

the machinations of boarding house runners, and the wily tricks of *soi-disant* emigrant agents, and finally after having learnt by costly experience the attractions of the "Empire City" they are forwarded to the railway stations from which they take the cars for the West. At Baltimore, however, the case is very different, there is no cartage of goods destined for the West, steamers come into port immediately alongside the piers and warehouses, erected by the Baltimore and Ohio Railroad Company, and the only expense of transfer is the movement from the ship to the railroad car over a platform of about 40 feet. This reduces the cost of handling to a minimum, and the fact is so thoroughly appreciated by the Western and Southern merchants, that while the gold coin duties in 1866 amounted to but little more than $4,000,000, they were upwards of $10,000,000, in 1873. Again look at the emigrant. Landing at the commodious piers alluded to, either from the North-German Lloyds' or Allan Steamers, he finds in a place secluded from all danger and annoyance a bureau-du-change, where English or German coin and notes can be exchanged for current funds, his baggage is passed by the custom house, and he takes the cars on the pier at the side of the steamer and is forwarded without cost and delay to such point in the West or North-west, as he may select. Accurate enquiry into the merits of the system pursued by the Baltimore and Ohio Railroad Company indicates that the advantages offered are duly appreciated. There are instances where steamers have recently come in with more than 800 emigrants, and each one realizing the comforts and conveniences of a port where he has been so well treated, becomes a *living* advertiser for the place where he first received such favorable impressions, he naturally indicates his preference for Baltimore as the objective point for all immigration from the "Faderland" or the "Ould Counthry." As an illustration of the progress which the foreign commerce of the port of Baltimore has made during the past eight years under these auspices, it may be noted that at the close of the war, the Baltimore and Ohio Railroad Company purchased from the government of the United States four steamships named respectively, Alleghany, Carroll, Somerset and Worcester. These vessels were found to have too limited a carrying capacity to insure remunerative returns and they were discontinued in 1870, but prior to their discontinuance in 1868, a contract had been entered into with the North German Lloyds' Line to establish a regular line between Baltimore, Southampton and Bremen. The two first built were named the "Baltimore" and the "Berlin," and experience demonstrated that the business was ample and remunerative from the inauguration of the interprise, in fact during the two succeeding years the accretions of traffic were so much in excess of 'their anticipations that in 1870, two splendid new steamers the "Leipsic" and the "Ohio" were added to the fleet, and this increase has been further supplemented since that date, that two additional new steamships, viz: the "Braunschweig" and the "Nurnberg" each of 3,000 tons burthen, and furnished with all the modern improvements, have been placed in the weekly line between Baltimore and Bremen via Southampton. The Liverpool steamship owners have also been sagacious enough to recognize the importance of the commerce which converges to Baltimore, and the Managers of the Allan Line, after a brief (and it is believed unsatisfactory) experiment of Norfolk as a shipping point have within the past two years placed nine large and first-class steamers on the route between Baltimore and Liverpool. Definite importance has thus been attached to the commerce of Baltimore, its geographical advantages for controlling the import and export trade of a large section of the United States, have been unequivocally demonstrated, and its merchants will be strangely recreant to the principles and aspirations by which they have been hitherto animated unless they make still further developments and attract to their city a traffic which will keep a fleet of at least thirty steamers constantly occupied. These suggestions are not chimerical because if results have been attained during eight years similar to what have been alluded to above with Western and South-western connections imperfectly and inadequately developed and with the most prolific portion of our common country, paralyzed by the prostration incident to the war and the *reconstruction* policy of the Federal Government, what may reasonably be anticipated when the South returns to its normal condition of prosperity and affluence, when its merchants and planters may be found congregating as of old to the marts at the head of the Chesapeake Bay, and when through the opening and successful operation of the Baltimore, Pittsburg

and Chicago Railroad, Baltimore can for the first time demand
without let or hindrance, and by an independent line entirely
under its own control, a fair and impartial representation in the
produce markets of Chicago and the North-west. Special allusion
is made to this point, because in all appearances an entire revolution will be produced in the carrying trade between the North-
west and the Atlantic seaboard by the completion of the new line
to Chicago. Its cost will not be one-half that of other roads with
which it comes into competition and a lower rate of freight coincident with a reduced cost of construction must be the means of
diverting to Baltimore much traffic which has hitherto been directly tributary to New York, Philadelphia or Boston. Another
commercial advantage of Baltimore in connection with its European line of steamers is the cheapness of fuel. The coal from the
Cumberland basin is pronounced superior for steaming purposes,
to any except the South Wales coal, and of that only one seam is
we are creditably informed, superior. The supply of coal in
England is found to be annually diminishing, and its prices are
effected by exceptional causes, more especially by the eccentricities
of miner's unions, &c., hence it is almost regarded as certain, that
within a few years if not sooner, America will furnish for the
European steamers, the West Indies and South America much of
the fuel which has heretofore been derived from trans-Atlantic
ports. In this however we may have formed premature conclusions
and would only state the simple fact that if a steamer sailing from
Baltimore to Europe uses 800 tons of coal on the voyage, her
managers will save by coaling at that point as compared with
New York or Philadelphia, $24,000 or £4800 per annum. In
these days of exaggerated and abnormal competition for ocean
freights, and when all the steamship lines formerly embraced in
what was known as the North-Atlantic conference are claiming
that their expenses are very largely in excess of receipts, this item
of cheap and reliable fuel is one that cannot be disregarded, it is
one which must eventually influence other steamship lines to make
Baltimore their permanent port. Much more might be written
on this point and it might be shown, how the commerce between
the United States and South America could with proper management be almost entirely concentrated at Baltimore, in fact a large

proportion of the coffee shipments from Brazil for the West now
pass through Baltimore, and the traffic originating in the valley
of the Amazon, in Bolivia, the Argentine Republic and the United
States of Columbia must find its readiest and economical market
there; but enough arguments have been adduced to convince even
the most skeptical that the seaport situated at the head of the
Chesapeake, the boldest indentation on the Atlantic, with advantages of distances, lower port charges and more economical transportation facilities must attract the attention and intelligent action,
which its location demands. Years may elapse before all that the
anticipations formed as to the commercial future of Baltimore are
fully realized, but farmers, merchants, manufacturers, laborers
and mechanics are now studying *economy*, they are anxiously
looking over the net results, and as a consequence they will abandon ports where the cost of handling and transferring freight are
simply extravagant and exorbitant and they will avail themselves
of advantages and commercial facilities which cannot fail to add
to their financial permanence and stability.

The city of Baltimore is also admirably situated for manufacturing purposes, and here it may be stated that manufactures flourish
in a locality where there are cheap fuel, abundance of water, cheap
rents and a cheap market for all the necessaries—not the luxuries
of life. It is a well known fact that the water power on which the
manufacturers of the Eastern States heretofore relied, has, during
the past decade, failed to a very great extent, hence they have been
compelled in the majority of instances to use steam as a reliable
motor, and the necessary fuel has been procured at considerable
expense from Philadelphia, Baltimore and New York. The prices
of coal procured from Philadelphia and New York have constantly
fluctuated, hence manufacturers have always had an uncertain basis
to work on, the value of their productions was always contingent,
and the actual profit of their business could only be roughly estimated in advance. The disadvantages of such a condition of
affairs, will be readily appreciated by those who are conversant
with the competition of manufacturers, and with the small margin
on which they operate. A rise of $1.00 or 75 cts. per ton in the
price of fuel, seriously interferes with small profits; and hence it
is not extraordinary that young and enterprising capitalists from

all sections of the country have selected Baltimore and its suburbs
as a profitable place for investments, and have been attracted
thither by the allurements of cheap rent and cheap fuel. The
Baltimore and Ohio Railroad Company with their usual sagacity'
recognized this fact, and the annually increasing consumption of
coal, indicates that their efforts in the direction of a low and uniform rate on coal transportation have been successful, and that
their consistent policy to advance the prosperity of Baltimore, has
not been unrewarded. The following remarks are found in the
annual report for fiscal year ending Sept. 30th, 1872. "The
establishment and maintenance of low and uniform rates, enabling
consumers to rely upon their supplies being furnished throughout
the year at prices which will not be affected by changes in the
charge for transportation continue to cause a general and large
demand. The tariff of the Baltimore and Ohio Road for coal, has
continued summer and winter without alteration, for nearly five
years. The company has uniformly declined to enter into any
combinations to obtain advanced rates, and proposes to continue
this liberal and useful policy." It would not be surprising if Baltimore were to become within a few years, the chief manufacturing
point for all the cotton produced on the Atlantic seaboard, and
that through economy of operating, she was able to compete successfully, if not outrival these older established New England
cities, with which the manufacture of *domestics* has hitherto been
a specialty. Cheap rents also enter conjointly with cheap fuel into
the calculations of the manufacturer and the operative; and here
from her geographical location and the large extent of country
embraced within her limits, Baltimore can hold out advantages
superior to those of any other large city. The manufacturer, the
mechanic, and the artizan, can each procure the location which they
want, at a very moderate ground rent, extending over a long term
of years; and as a natural result the city is populated by a thrifty,
industrious laboring class, while its manufactories assume a character of stability and permanence which are rarely found wherever
the ground on which the buildings are erected is not owned in fee
simple. The report of the Canton Company for year ending
March 31st, 1874, substantiates very clearly the statements made
on this point, indicating that even at that distance from the heart

of the city, and in a time when all manufacturing industries were suffering from a depression incident to the panic of October, 1873, they leased on ground rents, and mostly in small lots, property whose ground rent if capitalised, would be equal to more than $320,000. Reference to the map accompanying this pamphlet, and to which the reader is referred as shewing accurately all the country within a radius of fifteen miles from the city of Baltimore, will prove that there is abundant water supply for all the manufactories which now exist or which hereafter may be erected in the city proper or in its vicinity. A cheap market was also assumed to be one of the attractions for a manufacturing population, and in this respect, Baltimore is candidly admitted to be superior to either New York or Philadelphia. Allusion is made here to the necessaries, not the luxuries of life. The following list of market rates as prevalent in the three cities above mentioned is herewith appended, as indicating very clearly the correctness of this position:

LIST OF MARKET RATES AT NEW YORK, PHILADELPHIA AND BALTIMORE.

CHARACTER OF ARTICLE.	AT NEW YORK. October 17th, 1874.	AT PHILADELPHIA. October 17th, 1874.	AT BALTIMORE. October 15th, 1874.
Porter House Steaks	35 cts. per lb.	25 to 30 cts. per lb.	25 cts. per lb.
Sirloin Steaks	20 to 28 cts. per lb.	25 to 30 cts. per lb.	20 cts. per lb.
Round Steaks	22 cts. per lb.	20 cts. per lb.	16 cts. per lb.
Rib Roast of Beef	26 cts. per lb.	20 to 25 cts. per lb.	18 to 20 cts. per lb.
Corned Beef	12 to 16 cts. per lb.	18 cts. per lb.	10 to 12 cts. per lb.
Roast Veal	20 to 22 cts. per lb.	16 cts. per lb.	15 cts. per lb.
Veal Cutlets	35 cts. per lb.	20 to 25 cts. per lb.	20 cts. per lb.
Roast Mutton	18 to 20 cts. per lb.	16 to 18 cts. per lb.	15 cts. per lb.
Mutton Chops	25 cts. per lb.	16 to 20 cts. per lb.	15 to 18 cts. per lb.
Roast Pork	14 cts. per lb.	15 to 16 cts. per lb.	15 cts. per lb.
Pork Steaks	14 cts. per lb.	15 to 16 cts. per lb.	15 cts. per lb.
Corned Pork	14 cts. per lb.	16 cts. per lb.	15 cts. per lb.
Breakfast Bacon	16 cts. per lb.	16 cts. per lb.	16 cts. per lb.
Butter	38 to 50 cts. per lb.	40 to 60 cts. per lb.	30 to 35 cts. per lb.
Eggs	37½ cts. per dozen.	30 cts. per doz,	28 to 30 cts. per doz.
Chickens	20 to 26 cts. per lb.	18 to 20 cts. per lb.	$3 to $5 per doz.
Potatoes	60 cts. per peck.	$1.40 per bushel.	30 to 40 cts. per peck
Sweet Potatoes	80 cts. per peck.	$1.50 per bushel.	40 cts, per peck.
Onions	80 cts. per peck.	60 cts. per peck.	60 cts. to $1 per peck.
Pumpkins	40 to 60 cts. each.	40 to 60 cts. each.	20 to 30 cts. each.
Turnips	50 cts. per peck.	75 cts. per bushel.	40 cts. per peck.
Tomatoes	40 cts. per peck.	75 cts. 2 peck bask't.	30 cts. per peck.
Cauliflowers	35 to 45 cts. per head	30 to 40 cts. per head	25 to 40 cts. per head
Apples	50 cts. per peck.	50 cts. per peck.	25 to 40 cts. per peck.
Grapes	5 to 10 cts. per lb.	5 to 10 cts. per lb.	5 to 10 cts. per lb.
Halibut	18 to 20 cts. per lb.	16 cts. per lb.	20 cts. per lb.
Lard	17 cts. per lb.	12 to 16 cts. per lb.	16 cts. per lb.
Flour	$5 to $5.75 per bbl.	$6 to $7 per bbl.	$4 to $5 per bbl.

Some idea of the manufactures of Maryland and their varied character may be formed from the following Table of Statistics, compiled from the U. S. Census Returns of 1870, which show that out of products amounting in the aggregate to $76,593,613, $59,219,933 were credited to Baltimore County and City:

Character of Manufacture.	No. of Establishments.	Employés.	Capital.	Wages.	Materials.	Products.
Acid—Sulphuric	1	35	$ 50,000	$ 21.500	$ 19,880	$ 58,350
Agricultural Implements	34	295	281,300	117,311	276.257	549.085
Awnings and Tents	3	14	5,400	4,200	23,900	37.200
Babbitt—metal and solder	2	8	75 300	2,500	50,800	57.690
Bags—paper and others	5	116	109,540	31,950	441,260	583,275
Banners, Flags, &c	3	25	29,500	10,380	24,600	42,500
Bark—ground	2	3	14,100	400	3,850	5,200
Baskets	1	12	700	2,000	937	4,500
Belting and Hose, (leather,)	4	16	28,500	7,700	19,060	38,512
Billiard Tables, &c	1	2	1,500	720	4,500	7.260
Blacking	1	3	1,000	720	1,720	9,000
Blacksmithing	615	1,329	245,876	171,157	237,683	782,165
Bleaching and Dyeing	15	33	4,250	8,918	6,490	32,090
Book-Binding	16	125	41,350	38,994	30,470	80,374
Boot and Shoe Findings	3	40	2.900	5,760	13,672	28.200
Boots and Shoes	812	3,228	767,105	920,145	1,341,163	3,111,076
Bottling	4	28	9,000	10.720	21,500	44,100
Boxes of all sorts	28	300	84,500	78,226	193,108	358,359
Brass Founding, &c	5	136	58,000	61,559	85,676	255,435
Bread, Crackers, &c	159	488	374,195	124,168	765.914	1,220,399
Brick	73	2,051	1,063,300	588,283	256.963	1,191,515
Brooms and Wisps	11	218	70 575	35,836	183,784	277,938
Brushes	4	31	11,500	8,300	15,513	29,000
Butchering	21	77	39,020	17,785	108,943	164,465
Carpentering and Building	191	820	269,430	275,947	1,025 667	1,699,502
Carpets	20	50	22,950	10,230	26,755	46,507
Carriage Trimmings	3	48	21,500	11,000	16,000	40,000
Carriages and Wagons	133	681	297,650	227,170	219,132	667,757
Cement	2	76	32,000	26,000	47,760	79,500
Charcoal	3	75	74,000	30,500	19,300	75,480
Chromos and Lithographs	5	63	69,050	31,100	22,400	78,000
Cider	3	6	1,150	678	2,600	4,100
Clothing	323	7,453	2,284,825	1,135,426	3,785,993	5,970 713
Coal Oil—rectified	8	55	198,000	23,120	516,171	617.389
Coffee and Spices	4	29	139,600	16,423	312,570	360,535
Coffins	30	66	23,350	10.708	21,825	64,600
Confectionery	52	279	249,585	73,450	475,704	733,431
Cooperage	88	745	290,454	275,217	415,037	873.782
Copper—milled and smelted	1	127	800,000	96,500	1,098,665	1,016,500
" " rolled	1	7	40,000	3 000	6,000	10,000
Coppersmithing	7	30	11,700	9,750	20,701	52,625
Cordage and Twine	7	46	22,150	8,331	24,862	44,645
Cordials and Syrups	2	16	29,000	6,300	15,574	27,996
Cosmetics	1	2	300	350	450	1,700
Costumes	1	3	4,000	450	1,000	1,500
Cotton Goods	22	2,860	2,734,250	671,933	3,400,426	4,852,8 8
Curled Hair	1	14	13,912	3,222	13,086	24,392
Cutlery—edged tools	5	8	1 800	700	245	3,550
Dentistry—Mechanical	6	11	6,110	3,520	3,720	13.404
Drugs and Chemicals	4	64	178,500	21,075	60,520	137,851
Dye Woods, &c.—ground	1	75	150,000	33,103	104,386	187,400
Engraving and Stencils	7	35	22,375	13,275	7,080	33,500
Explosives and Fireworks	1	5	12.000	1,200	3,600	7,000
Fertilizers	15	126	438.800	44,827	522,886	632 352
Files	3	8	1,150	1,924	500	6,000
Flouring and Grist Mills	518	1,101	2,790,700	178,733	5,828,471	6,786,459
Fruits and Vegetables—canned	19	1,985	603 800	263,419	1,028,818	1,587,230
Furniture and Chairs	137	1,134	847.970	450,824	572,778	1,399,488

TABLE OF STATISTICS—*Continued.*

Character of Manufacture.	No. of Establishments.	Employés.	Capital.	Wages.	Material.	Products.
Furs—dressed	5	36	$ 68,500	$ 8,470	$ 31,630	$ 65,500
Gas	5	427	1,820,000	276,294	374,750	1,027,165
Gilding	1	4	300	1,184	2,860	5,300
Glass—cut, stained and window	4	145	145,700	109,600	88,300	246,400
Gloves and Mittens	6	68	6,450	2,750	7,978	15,300
Glue	1	1	500	425		750
Gold Leaf and Foil	2	12	3,200	3,964	12,980	21,000
Grease and Tallow	2	6	17,500	2,350	44,210	51,000
Gunsmithing	7	19	5,250	7,216	2,300	16,200
Hair Work	4	19	11,100	3,700	11,700	24,250
Hardware	9	31	8,000	12,437	9,786	38,717
Hats and Caps	15	54	16,000	13,082	29,502	57,266
Heating Apparatus	2	48	125,000	30,063	83,000	137,211
Hoop Skirts and Corsets	5	32	6,600	5,350	23,800	43,170
Hosiery	1	3	100		780	1,000
Hubs, Spokes, Felloes, Shafts, &c.	4	17	6,100	3,700	7,750	14,900
Husks—prepared	1	2	9,450	730	2,210	4,920
Instruments—Profes'l & Scientific.	7	37	32,800	12,600	5,420	30,750
Iron—forged and rolled	7	1,444	983,000	700,922	1,300,315	3,573,312
" Bolts, Nuts, Washers	2	16	28,000	5,500	40,800	65,000
" Railing, wrought	4	19	4,500	4,700	6,325	16,000
" pigs	14	859	2,005,000	255,941	1,286,881	2,143,089
" castings, not specified	37	632	736,635	189,305	416,648	835.024
" Stoves, Heaters, Hollow ware	6	63	47,500	26,500	36,722	93,070
Japanned Ware	2	7	900	1,350	1,900	4,200
Jewelry	8	46	52,950	25,082	36,500	52,700
Kaolin and ground earth	1	3	5,000	900	5,000	6,000
Kindling Wood	3	15	15,500	5,950	16,315	28,500
Lasts	2	2	400		250	2,100
Lead—Shot	1	12	42,000	2,550	59,715	62,354
Leather—tanned	69	281	792,430	94,180	926,406	1,265,388
" curried	50	163	238,145	53,131	494,075	623,308
" Morocco, tanned, &c.	4	84	64,000	39,200	74,880	163,000
" dressed skins	2	19	25,550	9,400	21,500	33,000
Lightning Rods	2	3	1,075	375	1,280	6,000
Lime	24	150	106,150	48,505	90,499	234,199
Liquors—distilled	8	63	220,700	30,228	381,236	889,261
" malt	32	208	583,500	103,644	510,492	665,743
Locksmithing, &c	14	47	7,800	12,750	8,224	33,700
Looking Glass and Picture Frames.	15	108	48,300	38,470	58,310	138,750
Lumber—planed	11	151	241,800	62,068	320,632	474,857
" sawed	391	1,245	1,055,600	259,551	674.858	1,501,471
Machinery, (not specified.)	22	343	372,700	175,744	223.646	581 391
" Steam Engines, &c	7	341	435,666	174,527	182.730	373,475
Malt	3	54	145,666	20,780	298,428	239,500
Marble and Stone Work, general	13	135	148.227	57,830	73,451	168,999
" Monuments & Tombstones.	24	274	228,250	126,467	143.896	375,597
Masonry—brick and stone	38	107	11,960	23,996	91.868	151,410
Matches	1	105	20,000	16,000	34.000	110,000
Meat—Pork, packed	1	25	125.000	14.000	526,200	595,000
Meters—Gas	1	25	50,000	11,000	15,000	45,000
Millinery	31	72	18,275	6,954	27,455	54,719
Millstones	2	22	23,400	11,537	14,700	35,530
Millwrighting	4	14	1,400	2,200	5,450	10.750
Mineral and Soda Waters	7	25	23,300	6,071	8,178	34,380
Molasses and Syrup, (principally Sorghums)	6	32	3.900	1,895	4,050	8.230
Molasses Sugar—refined	4	437	958,000	108,551	6,394,569	7,007,857
Musical Instruments and Materials (not specified.)	9	384	594,000	249,348	316,570	674.600
Nets—Fish and Seine	1	26	100	375	750	2,000
Oars	3	10	3,800	1,760	9,620	16,130

TABLE OF STATISTICS—*Continued.*

CHARACTER OF MANUFACTURE.	No. of Establishments.	Employés.	Capital.	Wages.	Materials.	Products.
Oils—Vegetable, (not specified,).....	2	43	$ 145 000	$ 18,900	$ 371,400	$ 478.125
Oysters and Fish—canned	13	1,531	553,300	256,719	922,402	1,418,200
Painting...	50	181	23,450	57,314	70,004	191,435
Paints—(not specified,)....................	2	41	65,000	20,000	300,000	387,500
" lead and zinc......................	3	69	375,000	44,500	465,148	640,000
Paper—(not specified,).....................	2	9	6,500	4,060	16,725	32,910
" Printing...............	12	266	1,121,800	86,116	445,498	823,000
" Wrapping........................	12	63	78,700	20,413	43,208	92,800
Paperhanging.................................	6	36	48,500	12.984	41,838	65,784
Patent Medicines and Compounds..	14	97	201,350	27,333	131,850	282,250
Patterns and Models.....................	2	6	900	1,900	360	3,950
Pencils and Pens—Gold................	1	2	500	156	600	1,500
Photographs..................................	28	95	83,825	27,711	25,770	125,981
Pipes—Tobacco	2	24	2,100	9,080	2,774	18,900
Plaster—ground	5	12	13,950	1,973	11,788	18,300
Plastering.....................................	12	45	9,380	15,528	20,060	50,450
Plated Ware..................................	4	26	20,800	9,015	10,700	28,500
Plumbing and Gasfitting..................	27	163	113,700	66,006	232,806	408,107
Pocket-Books.................................	4	7	1,500	1,000	1,400	4,400
Printing and Publishing—(not specified,).....................................	40	665	826,800	404,305	334,556	1,179,928
Printing and Publishing—Job........	27	200	181,350	101,947	165,048	391,521
Pumps..	16	51	24,125	11,602	19 391	47,803
Roofing Materials..........................	2	23	65,000	14,000	44 237	80,853
Saddlery and Harness......................	135	420	207,385	109,806	267,007	539,033
Safes, Doors and Vaults—Fire-Proof.	1	13	20,000	9,000	6,000	35,000
Sails..	7	37	15,000	12,153	48 821	76,827
Sash, Doors and Blinds..................	17	262	282,425	149,014	214,284	419,506
Saws..	3	14	4,000	3,116	4,092	12,000
Scales and Balances......................	2	8	22,000	3,250	36 535	58,000
Sewing Machines............................	5	12	1,405	4,500	1,350	7,300
Ship Building, Repairing and Ship Materials.....................................	31	313	172,500	116,836	126,723	357,404
Showcases	2	10	4 500	3,736	10,800	18,300
Silver Ware...................................	4	43	82,300	32,400	25 940	68,000
Small Beer.....................................	1	6	10,000	3,000	6 000	14,000
Soap and Candles............................	13	99	230,050	38,392	315 972	521,439
Soapstone Goods............................	1	24	100,000	11,000	10,075	26,150
Stereotyping and Electrotyping......	1	16	6,000	3,200	5,747	10,000
Stone and Earthen Ware..................	21	160	105,590	57,401	32 871	143,114
Sumac—ground..............................	1	8	15,000	4,160	15,250	20,400
Tin, Copper and Sheet Iron Ware...	183	956	663,500	318,742	901,901	1,634 009
Tobacco and Cigars.........................	2	7	4,500	3,236	2,565	9,500
" Chewing, Smoking and Snuffing...................................	13	346	496,400	82,046	251,911	653,760
Tobacco—Cigars.............................	269	1,034	409,100	304,502	409,803	1,108,988
Toys..	1	11	5,000	1,200	2,000	5,000
Trunks, Valises and Satchels...........	13	76	45,250	27,228	55,084	114,100
Trusses, Bandages and Supporters.	3	5	1,450	100	1,013	5,250
Type Founding...............................	1	16	25,000	7,500	1,780	12,000
Umbrellas and Canes......................	5	10	4,000	1,820	3,600	10,700
Upholstery.....................................	13	68	100,650	23,515	71,395	139,400
Vinegar...	4	13	21,200	3,672	73,810	89,000
Watch and Clock Repairing............	40	86	32,775	22,781	8,487	66,485
Wheelwrighting	257	522	115,735	69,869	90,151	311,581
Whips and Canes	3	40	14,000	13,000	28,200	55,100
Willow Ware and Rustic Ornam'ts..	10	99	9,350	10,050	26 058	48,450
Wire Work.....................................	6	122	28,200	26,120	57,500	113,100
Wood Brackets,Mouldings &Scrolls	4	127	169,000	77,411	56,500	154,580
" turned and carved.............	6	76	20,000	34,170	5,700	75,882
Wool-Carding and Cloth-Dressing...	4	19	6,500	2,280	20,160	38,310
Woolen Goods.................................	28	309	198,045	79,739	214,369	390,036

From the statistics above given, and which are the latest officially published, it will be readily seen that Baltimore does not possess one, but many sources of manufacture; prominent among these may be noted:

(*1.*) *The Oyster, Fruit and Vegetable packing,* in which are employed more than 25,000 people distributed among one hundred or more establishments. As an indication of the magnitude of this traffic it may be noted that during the season, 50,000 cans of raw oysters are put up daily by a single house, and 30,000 cans of cooked oysters by another. During the time when oysters are not in season, the hands are employed in canning fruit and vegetables to be shipped to Europe and the western markets. Large lime kilns are in many instances owned by the packing establishments, and it is stated that one firm alone burns 20,000 bushels into lime every four days; also, that the manufacture of 600,000 bushels of lime in a year, does not dispose of the accumulations, and the removal of a large quantity has to be paid for annually.

(*2.*) *Whiskey.* The Maryland whiskey generally, and more especially that produced in the city of Baltimore and its vicinity, has, for years past, gained an enviable reputation and is in extensive demand throughout the South, also in the Eastern and New England states. This whiskey is made out of pure rye, and the capital invested in its manufacture amounts to $3,000,000; but the profits resulting from the business are large, and a considerable increase of trade may be reasonably anticipated in the future.

(*3.*) *Grain.* Baltimore has always been from the first years of its existence as a town, the grain market of this section, and during the Peninsular war large quantities were annually shipped to Spain and other European ports. Great care has always been taken in the manufacture of flour at this point, and it has always commanded a high price in the West Indian and other tropical markets. An unusual stimulus has been given to the grain and flour trade since the close of the civil war, by the construction of elevators and by economical transportation to Europe. These facilities will, in all probability, be further increased as the Northwest appreciates more fully the advantages of Baltimore as a shipping point. It may be noted that a very rigid inspection of all grain and corn shipped from Baltimore is practised under the

3

auspices of the Corn and Flour Exchange, hence cargoes shipped from this port maintain a deservedly high reputation. The census returns for 1870, which have doubtless been subsequently increased about 25 per cent., indicate that in the State of Maryland there were at that time 518 flouring and grist mills, employing 1,101 hands, with a capital invested of $2,790,700 and with products amounting in the aggregate to $6,786,459.

(*4.*) *Shoe and Leather Trade.* The manufacture and sale of shoes and leather in Baltimore is very large, amounting in some years to more than $18,000,000. The traffic in these articles was formerly confined to New England and Philadelphia, but during late years the trade of Baltimore with the South in these articles has increased beyond all expectations; and from present appearances the shoe trade must become one of its leading industries.

(*5.*) *Cotton.* The census returns for 1870 show that cotton manufactures are rapidly becoming a specialty in Baltimore. The greater portion of the cotton now brought to this port is produced in North and South Carolina, but the goods manufactured out of cotton, although amounting in value to nearly $5,000,000, form but a small proportion of the aggregate cotton shipments, which are increasing annually from all sections of the country in consequence of the increased facilities for railroad transportation and the advantages offered to brokers, by the establishment of the "Baltimore Cotton Warehouse Company." There is apparently no reason why Baltimore should not become one of the leading cotton markets of the world; and it is certain that cheap rents and cheap fuel will eventually attract much New England capital; in fact, Baltimore must become a strong and in all probability successful rival of Fall River, Lowell and other well-known New England manufacturing centres. At present the cotton mills of Baltimore are more particularly interested in the manufacture of "Cotton Duck."

(*6.*) *Iron.* The importance of this trade cannot be ignored, and it ranks high among the specialties of Baltimore. There are rolling mills, furnaces, etc., in which nearly one million and a half dollars have been invested, and the products of which, according to the last census returns, amounted to nearly $3,000,000. A large increase in this business however, has taken place within the past three years; and the facilities for procuring iron ore, cheap fuel and an abundant water supply are evidently appreciated. It may be noted in this connection, that rails or plates manufactured by the Abbott Iron Company have a national reputation, and the further development of similar first class industries will be co-incident with the gradual extension and growth of the city. In connection with this subject of iron manufacture it may be appropriately noted that the construction of iron bridges is carried on extensively in Baltimore by the "Baltimore Bridge Company," and by the "Patapsco Bridge and Iron Works." The last named company has executed considerable work in Cuba and Mexico, also in North Carolina at Wilmington; while the former has attained a world-wide reputation by the construction of the Rock Island Bridge across the Mississippi; of the St. Charles Bridge across the Missouri River, on the line of the St. Louis, Kansas City and Northern Railroad, and of the Varrugas Viaduct for the Lima and Oroya Railroad in Peru, 252 feet high. The superior quality of the iron manufactured in Baltimore, gives these companies unusual facilities for executing large contracts, and a large force of laborers is constantly employed.

(7.) *Petroleum.* The trade in this article at Baltimore, and the establishment of refineries, is increasing rapidly, but not in a ratio corresponding to the facilities which can be furnished for its manufacture and export. It is claimed that the Baltimore refineries could furnish oil cheaper than those of Philadelphia, did no discrimination exist against the first named city, in the transportation of the crude article from the Pennsylvania Oil Wells. The following statistics indicate the growth of the business for two years:

Receipts of Crude and Refined Oil.	1872.	1873.	Increase.
Per Northern Central Railway	93,397 bbls	180,590 bbls	87,193
" Baltimore and Ohio Railroad	53,977 "	66,396 "	12,419
Total	147,374 "	246,986 "	99,612

The capacity of the refineries in Baltimore, is 475,000 barrels annually; and this capacity could be increased indefinitely, provided that an independent line were constructed from Hagerstown, on the Western Maryland Railroad, to the coal oil regions of Pennsylvania. During the year 1873, twenty-nine cargoes cleared from

Baltimore for foreign ports, composed in the aggregate of 54,464 barrels refined; 11,951 barrels lubricating and 5,268 barrels naptha, making the total exports, 71,683 barrels, equal to 3,189,850 gallons, which, with the crude oil added, make the aggregate of foreign shipments 3,470,955 gallons, against 1,972,258 gallons in 1872. Within six years the trade in petroleum has increased more than 350 per cent., viz: from 988,236 gallons in 1869, to 3,470,995 in 1873. Succeeding annual reports will in all probability show a still further percentage of increase.

(*8.*) *Bricks.* Although when the first brick house, contiguous to Baltimore, was built, the bricks were imported from England; the inhabitants of the city soon discovered the value of the clay in its vicinity for manufacturing bricks of a very superior quality. Large sums of money amounting at this date to nearly $1,000,000, are invested in the business. More than 2,000 laborers are constantly employed, and 25,000 tons of coal, together with 2,000 cords of wood are consumed in brick production. Among the different kinds of brick manufactured here, the Baltimore *pressed brick* stands preëminent, and its superiority is so thoroughly recognized that it is shipped extensively to all the seaport towns lying south of Baltimore, and also largely to New England and Boston. In the neighborhood of the city a superior fire brick clay is found, and also clay which is adapted for stoneware, pottery and terra cotta ware. The pottery branch of this trade requires further development and the introduction of additional capital; because there is no reason why the manufacture of such wares here should not surpass that of Trenton, and monopolise the trade of the South and West, provided that the necessary site for such establishments can be procured at a reasonable price, and in proximity to rail and water transportation.

(*9.*) *Furniture.* This is another important industry of Baltimore, in which according to the latest returns accessible, more than two thousand hands are employed, while a capital of more than $1,500,000 is invested in the business. The annual sales now amount to $3,000,000, being nearly double of what they were when the census returns were taken in 1870. Black walnut is procured at a comparatively low rate from Indiana, while the forests of West Virginia furnish an inexhaustible supply of poplar, and superior yellow pine can be obtained in large quantities *via* Chesapeake Bay, from the forests of lower Maryland, Virginia and North Carolina. This branch of manufacture is increasing so rapidly, that to all appearances Baltimore will become the furniture emporium, not merely for Maryland proper, but for the South and West, for South America and the West Indies.

(*10.*). *Tobacco.* Last but not least among the varied manufactures for which Baltimore stands preëminent, is that of Chewing Tobacco, Smoking Tobacco, Cigars and Snuff. The brands manufactured in Baltimore have a very extensive sale both here and in Europe, and a large number of skilled laborers are constantly employed in the various factories. The export trade in leaf tobacco is one of the prominent features of business in Baltimore, and has been so since the first foundation of the city. There is a very rigid inspection of tobacco grown in Maryland, and the knowledge of this fact attracts large orders from foreign manufacturers to Baltimore. Heavy shipments are also made from Ohio, Virginia and Kentucky, through Baltimore. The Maryland and Ohio tobacco exported during the year ending December 31st, 1873, aggregated 51,652 hogsheads, of which 40,000 hogsheads were for Bremen, Rotterdam and France.

Allusion might be apprroriately made here to the manufacture of pianos, tin and glass ware of all descriptions, fertilizers, chemicals and many other articles for home use and for export; but enough has been written to show that the city of Baltimore possesses within herself all the resources for becoming a large manufacturing centre: and when it is considered that these resources are supplemented by cheap fuel, cheap rents, economical transportation and abundant labor, the reader will see that there is no reason why, by the attraction to itself of foreign capital for investment in these varied industries, Baltimore should not become, as it is designed by nature, the great commercial and manufacturing centre for the West and South.

The following statistics, relative to the imports for three years, ending December 31st, 1873, and the receipts of various articles by the Baltimore and Ohio and Northern Central Railroads during two years, will doubtless prove interesting and instructive:

IMPORTS.

Comparative Table of Imports and Receipts of Principal Articles for Three Years, ending December 31st, 1873 :

ARTICLES.	1873.	1872.	1871.
Coffee—Rio, bags	380,449	372,895	566,995
Cocoanuts, M	1,129	1.250	1.500
Cotton, bales	116,578	113,367	110,637
Flour, barrels	1,312,612	1,175,967	1,123,028
Corn, bushels	8,330,449	9,045,465	5,735,921
Wheat, bushels	2,810,917	2,456,100	4,076,017
Oats, bushels	1,255,072	1,959,061	1,833,469
Rye, bushels	100,519	90,938	88,956
Mackerel, barrels	17,314	15,690	26,202
Herring, barrels	20,767	24,715	36,755
Guano, tons	10,857	3,007	12,885
Lemons, boxes	26,281	21,732	28,102
Oranges, boxes	54,291	50,528	68,949
Raisins, boxes	45,734	91,229	64,652
Hides, No.	150,749	175,000	158,528
Iron, bars	9,507	10,940	32,600
Pig Iron, tons	3,261	5,418	7,250
Railroad bars	49,602	103,486	108,970
Iron, bundles	....	2,480	16,586
Molasses, hogsheads	30,718	25,006	28,158
Sugar, bags	31,827	36,600	49,129
Sugar, hogsheads	127,282	116,961	126,619
Sugar, boxes	65,107	79,188	55,044
Rice, tierces	17,228	11,082	11,397
Rice, bags	22,781	25,618	15,873
Salt, sacks	280,146	183,700	223,960
Salt, bushels	142,985	248,603	101,413
Spirits Turpentine, barrels	17,979	21,657	22,852
Rosin, barrels	80,346	80,020	79,352
Tar, &c., barrels	19,243	13,467	13,225
Tin Plates, boxes	184,822	190,511	135,316

BALTIMORE AND OHIO RAILROAD.

Comparative Statement of the Leading Commodities Received from the West by the Baltimore and Ohio Railroad, and Delivered to consignees at Baltimore for Eleven Months, ending November 30th, for 1872 and 1873 :

	1872.	1873.
Cotton, bales	2,970	12,556
Coal, tons	1,452,540	1,972,310
Flour, barrels	655,108	832,314
Wheat, bushels	263,800	601,100
Corn, bushels	4,107,643	4,725,393
Oil, barrels	53,977	66,396
Lumber, tons	18,398	39,715
Provisions, tons	27,870	25,817
Butter, tons	430	420

NORTHERN CENTRAL RAILWAY.

Receipts at Baltimore for the last Two Years Compared.

	1873.	1872.
Coal, tons	212,954	244,775
General Merchandise, tons	140,643	136,611
Flour, barrels	307,798	279,534
Grain, bushels	2,282,122	1,577,794
Live Stock, tons	14,149	12,915
Lime and Plaster, bushels	307,976	407,201
Pig Iron and Iron Ore, tons	28,035	27,612
Lumber, feet	30.153,703	23,855,458
Coal Oil, barrels	135,595	93,397
Butter, tons	1,306	1,560
Lard, &c., tons	3,801	1,503
Provisions, tons	24,014	20,622

The whole being equivalent to 645,575 net tons, as against 599,304 tons in the year 1872, an increase of 46,271 tons, and compared with 1871, an increase of 122,714 tons.

These figures speak volumes, as indicating that Baltimore is always able to furnish, for vessels importing, a return freight of the productions of this country. As long as the receipts of home products, for shipment and for manufacture into articles of export, continue to increase in the ratio above given, there is no fear of commerce declining or manufacturing being transferred to other cities.

(III.) SITUATION OF HARBOR.

THE harbor or basin proper of the city of Baltimore is situated on what is known as the north west branch of Patapsco River, and when the town, as it was then termed was first founded, and for more than fifty years afterwards, the water reached up to Exchange Place and Water street on the north and nearly to Charles street on the west, in fact the basin and dock at the present date do not occupy more than half the space, which was occupied by water in 1783. At that early stage however of the city's history, the deposits from what is known as Jones' Falls filled up the harbor or basin to some extent, and an impost of one penny and afterwards of two pence a ton was levied on all vessels, entering or clearing with the view of providing a fund for maintaining a proper navigable depth of water. Nearly ninety-four years have elapsed and still the basin as it is termed, although somewhat extensively curtailed of its pristine proportions remains in the same location, the inhabitants of Baltimore, adhere to it affectionately as an old friendly landmark, despite the noxious and mephitic vapors endangering the health of the city which rise periodically from its semi-stagnant pool, whenever some sailing craft or tug boat more adventurous or more deeply laden than its competitors stirs up the augean deposit. Dredging machines are kept continually at work, and the city is put to a heavy expense year by year to keep the basin and its docks and slips clear, but the sedimentary deposit is still increasing and the question arises, whether it would not be much better to fill up the present basin clear across from Fell's Point due west, and for the city authorities, after having thus reclaimed property which would be highly valuable for warehouse and storage purposes, to devote their attention to keeping the anchorage at Locust Point and its vicinity of sufficient depth for large vessels. It is well known that the sedimentary deposits of a stream like Jones' Falls, or discharge from sewers when coming

into contact with tide water are immediately precipitated and if
this rule holds good in instances where a basin or harbor has a
current able to clear out the settled matter how much more does
the rule hold good in the case of a basin like that of Baltimore
city where there is little if any current and where the mean rise
and fall of the tide does not in any case exceed two feet. A Baltimore newspaper, "the Gazette of September 22d, 1874," alludes
in very strong terms and pointed language to the present condition of the harbor, and says: "Even while we were celebrating so
jubilantly the completion of our ship channel, a ship drawing but
nineteen feet six inches of water was lying within biscuit toss of the
great elevator of the Baltimore and Ohio Railroad hard and fast
aground. A few days later two ships laden with guano were also
hard and fast aground parallel with the elevator. They were only
gotten off after a detention of several days. Similar instances have
occurred with other vessels which are fresh in the memory of our
merchants. Forty years ago vessels drawing twenty feet of water
found no difficulty in coming up to O'Donnell's, Gibson's, Belt's,
Corner's or Tenant's wharves or to Locust Point. There was
then eighteen feet of water at the mouth of the Falls and from
fifteen to eighteen feet from the Falls to the head of the Basin.
At present at the wharves above mentioned there is scarcely sixteen feet of water, and at Locust Point except where dredging has
been recently done not much over twelve feet. At the mouth of
the Falls there is not much more than nine feet whilst commencing
nearly opposite the mouth there is a bar which runs in an easterly
direction nearly to Locust Point, which has been caused by the
sediment flowing from the Falls to the harbor. The water on
this bar is but from five to ten feet deep at mid tide, leaving but a
narrow channel for vessels to pass to and from the inner harbor or
what is termed the Basin. At the coal wharves at Locust Point
it has been found necessary to dredge channels in order to get the
coal vessels out, the draft of these vessels ranging generally from
ten to eighteen feet, very few, however, drawing eighteen feet of
water." These facts as given in a Baltimore paper are quoted
with the view of demonstrating that the city of Baltimore in batt-
ling to retain its hold upon European and internal commerce has
certain natural difficulties to contend with and overcome at a great

cost; *i. e.*, if it is its desire to adhere to old land-marks and not adopt a course which common prudence and adherence to the ordinary rules of hygiene would dictate.

Years ago when sailing vessels were in vogue, objections might have been urged against the distance from the port to the open sea; and it may be presumed, that beating up Chesapeake Bay for nearly 200 miles in the face of head winds, was not at that date a very pleasant experience; but steam, and its general introduction on all large sea-going vessels, have entirely obviated these difficulties, and the question of short inland communication is the principal point to be settled in all problems of through transportation. It might, however, be urged that Locust Point and other deep-water frontages on the west side of Chesapeake Bay will, in all cases, command the preference as shipping points for European and Eastern or Southern ports, in view of the fact that wharf-frontage on that side of the Bay is not exposed to the wind and sea as that on the east side. The railroads also converging to Baltimore from the South and West, have their present termini on the west side of the city. In almost all large cities, the tide of improvement generally sets " Westward." There is no reason why in the case of Baltimore there should be an exception to this recognized idea. Locust Point and the western wharves of the North-Western Branch may not be able to accommodate the rapidly increasing commerce of Baltimore, but there are other western deep-water harbors, just as accessible to Baltimore as Hoboken and Staten Island are to New York; and the same energy and consistency of purpose, which has enabled Baltimoreans to utilize fully the short line geographical advantages of their location in connection with the West and Southwest, and by tunnelling mountains and bridging immense rivers, to make a trunk road over which the varied products of the interior should radiate to the city of their choice, will lead them to overcome what may be correctly regarded as minor obstacles, and to adapt their harbor facilities not merely to the requirements of a section of our common country, but to the progressive requirements, if necessary, of a whole continent.

(IV.) RAILROAD CONNECTIONS.

ALLUSION has been made in previous sections of this pamphlet, to the fact that Baltimore is, from its peculiar geographical location, at the nearest accessible tide-water on the Atlantic seaboard, and from its short rail line advantages, destined to become the great shipping point and commercial emporium for the West and South. It has been also shown how the commerce of the city has increased during the past eight years, since the establishment of a regular European line of steamers; and how, under the combined influences of cheap fuel, cheap rents, abundant water supply and a cheap market, an additional stimulus has been given to all manufacturing industries. The growth of the city proper has been traced from the time when in 1729 its area was limited to sixty acres, down to the present date, when the population is more than 320,000; and when the assessed, not real, value of property is more than $300,000,000. It will be appropriate now to trace out in detail the various railroad connections of Baltimore, and indicate to the reader who may, perchance, not be thoroughly conversant with these facts, that the *short* rail line advantages claimed for the city are not exaggerated; and it may be advisable in this connection to treat of the various railroad lines categorically, in accordance with the importance of each, and the magnitude of its operations.

(1.) *Baltimore and Ohio Railroad.*

The *main* line extending from Baltimore to Wheeling, a distance of 379 miles; connects at Hagerstown Junction, 79 miles from Baltimore with the *Washington County Railroad*, running from that point to Hagerstown. The importance of this connection must be apparent on reference to the map, because at Hagerstown,

control is secured of a portion of the traffic originating in that
section of the country, which otherwise would be tributary to the
Cumberland Valley Railroad and would be naturally diverted to
Harrisburg or Philadelphia; at HARPER's FERRY connection is
made with what is now known as the *Harper's Ferry and Valley
Branch,* which consists of the "Winchester and Potomac," "Winchester and Strasburg," "Manassas Gap Extension" and "Valley
of Virginia" Railroads. The first named road was leased by the
Baltimore and Ohio Railroad Company in 1867, for a period of
twenty years, at an annual rental of $27,000. The second named
was leased in 1870, for a term of seventeen years at a rental equivalent to seven per cent. annually on its capital stock. The third
named, extending from Strasburg to Harrisonburg was leased
during the autumn or early winter of 1873, while the last named,
as being a road under process of construction to Salem, and one in
which the city of Baltimore and the Baltimore and Ohio Railroad
Company have a large pecuniary interest, is operated in connection
with the road from Harper's Ferry to Harrisonburg; but on what
terms is not known, presumably, however, at cost. This connection at Harper's Ferry, renders the traffic originating at present in
the Shenandoah Valley, tributary to the market of Baltimore;
and its present commercial value should not be under estimated, but
its prospective importance when the Valley Railroad is completed
to Salem on the line of the Atlantic, Mississippi and Ohio Railroad, is very great; inasmuch, as it must cause a large proportion
of the traffic of the Chesapeake and Ohio, and the Atlantic, Mississippi and Ohio Railroads to converge to Baltimore instead of
Richmond and Norfolk.

Relative to this *Valley Railroad,* it may be noted that it passes
through the centre of the great valley of Virginia, a district unsurpassed in fertility of soil and in mineral and agricultural wealth,
and in the thrift and energy of its rapidly increasing population.
It will command from its opening a heavy local traffic and must
prove a remunerative investment to its originators. It may be
noted in this connection that the Baltimore and Ohio Railroad
Company in pursuance of its consistent policy to build up the
commercial interests of the city of Baltimore, has been compelled
to lend a helping hand to the new railroads which were originated

to develop the trade of the South, and thus to neutralize the persistent efforts of the Southern Security Company to divert the traffic either to Philadelphia or New York. That these efforts were mainly dictated by a determined opposition on the part of the Pennsylvania Railroad Company, to the existing management of the Baltimore and Ohio Railroad Company is well known; also that advantages were taken of the impecunious condition of some prominent southern railroads, to obtain their control. Unexpected circumstances have conspired to defeat in some measure a programme, which aimed practically at shutting off the Baltimore and Ohio Railroad from its geographical alliances with the South and the Gulf states; so that Baltimore can now renew its business relations with that section of country, by more than one line of railroad; meanwhile, it is satisfactory to know that the *Valley Route* via Harper's Ferry, will form an important, and in all probability the shortest available all rail line between Baltimore, New Orleans, Mobile and other southern cities. At *Cumberland*, connection is made with the PITTSBURG, WASHINGTON AND BALTIMORE RAILWAY, (Connellsville route,) and a short line established between the great laboratory of the United States and tide-water. It may be noted here that prior to the construction of railroads, the whole business of the section of country lying between Pittsburg and Cumberland was transacted at Baltimore. The opening of the Pennsylvania Railroad from Pittsburg to Philadelphia caused the diversion of traffic into a new channel, but every effort is now being made by the merchants of Baltimore to regain their lost supremacy, and if direct connection can once be established with the oil regions of Pennsylvania, by the construction of an independent road from near Pittsburg on the Connellsville route to a junction with the Alleghany Valley Railway, a stimulus would at once be given to the petroleum traffic at Baltimore which would contribute very largely to its commercial prominence. At *Piedmont*, by connection with the system of the Cumberland and Pennsylvania Railroad Company, Baltimore obtains control of a coal business whose volume is immense, and which is annually increasing. The direct and indirect advantages accruing to the city from this coal trade cannot be overestimated, and the day is not far distant when its magnitude can only be limited by the terminal facilities for handling it and guaranteeing its expeditious and economical shipment. At *Grafton* the North-west Virginia Railroad, now known as the *Parkersburg Division*, diverges from the main stem, and large expenditures have been made by the Baltimore and Ohio Railroad Company between Grafton and Parkersburg, with the view of perfecting their railroad connections with Cincinnati, Louisville, St. Louis, the Mississippi Valley and generally with the South-west. It is believed that $9,000,000 have been invested in reconstruction of the North-west Virginia Railroad, arching its twenty-three tunnels, and building the magnificent iron bridge over the Ohio river at Parkersburg, and that such outlay was fully justified is evidenced by the rapidly increasing business from the section of country which it taps, and the greater proportion of which converges to Baltimore as a market. With reference to the advantages to be derived by the city of Cincinnati from availing itself of the short line via Parkersburg, the President of the Baltimore and Ohio Railroad Company made the following appropriate remarks in 1870: "For Cincinnati the advantages are most palpable. The average distance in favor of Cincinnati in communication with Baltimore, as compared with New York, is 240 miles. Can it be possible that with such immense advantages, with unequaled piers and fire-proof warehouses furnished without charge for foreign steamships, with the cheapest and enormous facilities for transportation between the East and West—can it be possible that if Baltimore will but continue her vigor and enterprise, will furnish additional lines of steamships to Europe, that the business of all these vast regions will not be attracted through *their* interests to Baltimore instead of New York? Can be possible that when more than 200 miles of land transportation can be saved in the interests of the farmer and the consumer in the West, that this great advantage will not be availed of? The *Queen City* will yet reach its highest prosperity and command enlarged trade through the use of its *shortest and cheapest outlet to the ocean.* It could thus compete boldly and successfully with any Western city, and its situation in relation to the trade of great territories would be superior and impregnable. We said to her citizens, that Baltimore had long recognized the strength of Cincinnati, that the preceding administration, and for nearly twelve years the present administration, of the Baltimore and Ohio Railroad Company had continued to spend its capital, in all that period without net result, in constructing the *shortest* line between the cities, until upwards of $10,000,000 have been invested in the line from Grafton to Cincinnati; $8,000,000 have been expended in building the Parkersburg Branch; $1,000,000 for the bridge at Parkersburg to connect the Marietta and Cincinnati with the Baltimore and Ohio Road, and $1,000,0000 of aid has been extended to the Marietta and Cincinnati Company. Our conviction has been that this line must be perfected, its tunnels permanently arched, the bridge erected, and the entire line made first-class. Thus our millions have been expended. We still believe that our faith has not been misplaced, and that soon this splendid and shortest line will be adopted as the great highway for commerce and travel and prove the source of the greatest fruition to the communities interested." Again, in connection with Louisville, President Garrett remarked in the same tone in 1870: "In a short period Louisville can command this very improved, direct and economical route to the seaboard. The favorable comparative distances of which you can thus avail for your foreign and general commerce, are very remarkable. To Baltimore, the distance by this line through Cincinnati, is 696 miles. To New York, by the Ohio and Mississippi and New York and Erie, it is 987 miles—291 miles *further*. By the New York Central, 989 miles—making 293 miles greater distance; and by the Allentown route of the Pennsylvania Road, the distance is 155 miles greater." Again referring to St. Louis and the shortest route to the seaboard, the following statement was made: "Passing from Chicago to St. Louis, Baltimore reaches its own parallel, and affords for that great and progressive city, the *shortest* and most direct route to the seaboard. Through the Ohio and Mississippi Railroad to Cincinnati, and the Marietta and Cincinnati Road; thence the Baltimore and Ohio Road presents a line 210 miles less in distance to Baltimore, than the average distance by the three trunk lines used from St. Louis to New York. That city to maintain and increase her commerce, must avail of the vast advantages of this short route and of the economies of the port of Baltimore. St. Louis appreciates the necessity of close, improved and increased relations with Baltimore. Her leading and most thoughtful citizens express their anxiety to secure a cordial and effective alliance with Baltimore through its great highway—the Baltimore and Ohio Railroad." Turning again to the main stem, the reader will find that at *Benwood*, 375 miles from Baltimore, connection is made with the Central Ohio Railroad, by another magnificent iron bridge over the Ohio river, and at Wheeling, with the Cleveland and Pittsburg and Hempfield Railroads, the latter of which extending from Wheeling to Washington, Pa., is controlled and operated by the Baltimore and Ohio Railroad Company. The value of this connection at *Benwood*, and the importance of securing in the interests of Baltimore its independent control, was thoroughly appreciated and acted on by the Baltimore and Ohio Railroad Company at the close of the war, or rather in 1866, when a provisional contract was entered into for leasing the Central Ohio Railroad at a fixed percentage of earnings, with the understanding that such percentage should in no case be less than $160,000 per annum, a sum which was required to meet the annual interest on the bonded debt of the lessors and the annual contributions to the sinking fund. At the time of making this contract, which subsequently was shaped into a lease for twenty years, renewable indefinitely, the Baltimore and Ohio Railroad Company regarded the action as important in view of the connections for Cincinnati, Indianapolis, and other Western points made *via* Columbus. The anticipations then formed as to the prospective volume of traffic to be obtained at Columbus, have been somewhat interferred with by leases of roads west of that point in a rival interest, and at what is considered by many as an extravagant rental; but arrangements have been made for obtaining a large amount of the traffic converging to Indianapolis, over the Indianapolis, Cincinnati and Lafayette Railroad; and although the lease of the Central Ohio Railroad has not hitherto been a direct source of profit to the lessees, still advantages have accrued to the city of Baltimore from the lease, and these advantages will be much more appreciated when the line between Bellaire and Newark becomes an important link in the new short line route, over which the almost unlimited traffic from Chicago and the North-West will be carried. At *Newark*, thirty-three miles from Columbus, the western terminus of the railroad system *via* Bellaire, under the control of the Baltimore and Ohio Railroad Company, connection is made with the Sandusky, Mansfield and Newark, and the Newark, Somerset and Straitsville Railroads, both of which are leased and operated respectively as the *Lake Erie and Straitsville Divisions*. By control of the first named road from Newark to Sandusky, the lessees obtained a favorable line to the Lakes, and *via* Monroeville, an available route to Chicago, over the Lake Shore and Michigan Southern Railway, while by the latter access was gained to valuable, and from present appearances, highly remunerative coal fields. In consequence of heavy expenditures requisite to put the road between Newark and Sandusky in first class condition, the lease of this property has hitherto resulted in a loss, but a large amount of traffic originating in Chicago, Toledo, and other places, has been diverted to Baltimore as a market in consequence of this lease, and as the whole transportation of the new Baltimore, Pittsburg and Chicago Railroad, will, within a few weeks, be concentrated on ninety miles of the Lake Erie Division, viz: from Centreton to Newark, the lease of this property must result in profit during future years. Relative to this new road, from Centreton to Chicago, which is destined to have such an important bearing, not merely on the annual earnings of the Baltimore and Ohio Railroad and its leased lines, but, on the commercial future of the city of Baltimore, it may be appropriately noted here that it commences at Centreton, and thence follows a course about midway between the Lake Shore and Michigan Southern Railway, on the north, and the Pittsburg, Fort Wayne and Chicago Railway, on the south, through Ohio and Indiana, passing by the flourishing towns of Republic, Tiffin, Fostoria, New Baltimore, Deshler, Holgate, Defiance and Hicksville in Ohio and Auburn, Avilla, Albion, Milford, Syracuse and New Bremen in Indiana, the entire distance from the Lake Erie Division to Chicago being about 268 miles. In locating this line great care has been taken to secure low gradients and easy curvature, while as little deviation as possible is made from an *air line*. It is stated on good authority that the line as located does not vary more than $2\frac{6}{10}$ miles from the air line, and the maximum grade will not exceed 26 feet to the mile. With these advantages of easy grades,

4

an air line and curves of not less than one degree, and with liabili-
ties per mile, much less than those of other roads with which it
will be placed in competition, it would be unreasonable to suppose
that Baltimore will not be able to control through cheap rates and
expeditious transit a large traffic with the North-west, from which
it has hitherto been practically debarred through not having an
independent line to Chicago. The value of this new road will be
apparent from a resumé of the connections which it makes between
Centreton and Chicago. At *Tiffin* it connects with the Cincinnati, Sandusky and Cleveland and the Mansfield, Cold Water and
Lake Michigan Railroads; at *Fostoria* with the Lake Erie and
Louisville Railroad; at *Deshler* with the Cincinnati, Hamilton
and Dayton Railroad; at *Defiance* with the Ohio State Canal
and the Toledo, Wabash and Western Railway; at *Auburn* with
the Detroit, Eel River and Illinois and Fort Wayne, Jackson and
Saginaw Railroads; at *Avilla* with the Grand Rapids and Indiana Railroad; at *Milford* with the Cincinnati, Wabash and
Michigan Railroad; and at *Walkerton* with the Indianapolis,
Peru and Chicago Railroad, while connection will be made with
the Michigan Central and the Lake Shore and Michigan Southern
Railways some twenty miles east of Chicago. Special favorable
arrangements have been made with the Illinois Central Railroad
Company for joint use of track and depot ground at the Lake
Front in Chicago, and a tract of land covering forty acres has been
secured in South Chicago, on which it is proposed to erect the
largest shops yet erected in or out of Chicago. The engine house
will have stalls for thirty-two locomotives and there will be car
shops and work shops of all kinds to correspond. It may be noted
that the road from Centreton to Chicago will not cost when completed much more than $6,000,000 or an average of $23,000 per
mile, that the necessary funds have been furnished by the Baltimore and Ohio Railroad Company, and that there is no funded or
floating debt as a lien upon the road. Numerous pertinent facts
and figures might be given here as indicating the superior financial basis on which the system of railroads now controlled by the
Baltimore and Ohio Railroad Company in the interests of Baltimore rest as compared with that of its trunk line competitors, but
their perusal and study would doubtless weary the reader and

even thus much about western connections would not have been written had it not been deemed advisable in justice to the commerce of Baltimore to show how much her geographical advantages for becoming the port for Western produce will be increased by the completion of this new road. There will be three connections for Toledo, and the merchants will be able to avail themselves in consequence of low competitive rates. The whole section of country, drained by the Toledo, Wabash and Western Railway and its branches including Central and Southern Illinois and Missouri will become tributary to Baltimore by the new connection at *Defiance*, while from *Auburn* and from *Avilla* in connection with the vast forests of Michigan and Indiana can be built up a lumber trade which will equal if not surpass that of other rival cities. While however the Baltimore and Ohio Railroad Company has been closely occupied in perfecting its western connections, improvements in connection with the South have not been neglected, aid has been as stated above extended to the Valley Railroad, a new connection has been established at Alexandria with the Washington City, Virginia Midland and Great Southern Railway, the Metropolitan Branch Railroad shortening the distance between Washington and Baltimore and principal Western cities, has been completed, in fact every thing which was possible has been done to carry out the ideas by which the originators of the Baltimore and Ohio Railroad were animated, and to utilize geographical advantages of location by the establishment of a perfect railroad system which would render the business of the South-west and North-west tributary to Baltimore.

(2.) Northern Central Railway.

The railroad connections of Baltimore, via the Northern Central Railway, and its various tributaries, are, in some respects, as important as those of the Baltimore and Ohio Railroad, which have just been described in detail; but the policy of the managers controlling the line, via Harrisburg, has not been consistently directed to building up the commerce of the city, and they have labored under difficulties in the shape of terminal facilities which were not experienced by their competitors. At the same time, it cannot be denied that if the Northern Central Railway had been governed directly in the interest of Baltimore, it would have been instrumental in building up more varied industries ; in fact, while the allegiance of the Baltimore and Ohio Railroad is due to Baltimore only, that of the Northern Central Railway is divided between Philadelphia and Baltimore, and this state of affairs must continue so long as a controlling interest in the management is held by the Pennsylvania Railroad Company. Statements given in a previous portion of this pamphlet indicate the tonnage brought to Baltimore via the Northern Central Railway, and its varied character. The iron, marble, lime and granite of Baltimore county stand side by side in the yards with the anthracite coal of Pennsylvania, the coal oil of Venango county, the rich ores of Lake Superior, the lumber of Western New York, and the cereals or other products of the West and North-west, in brief, all the resources of this vast country are represented in the general traffic of the Northern Central Railway, and there is no doubt that a much larger representation could be secured, in the event of suitable arrangements being made for terminal facilities at tide-water. The line now owned by the Northern Central Railway Company was commenced a short time after the Baltimore and Ohio Railroad, under a charter granted to the "Baltimore and Susquehanna Railroad Company," and it was expected at that time that concurrent legislation would be obtained from the Legislature of Pennsylvania, but in this respect the original projectors of the road were disappointed, and it was not until 1832 that authority was given to incorporate the "York and Maryland Railroad," said legislation having been further supplemented in 1846 by a charter incorporating the "York and Cumberland Railroad Company," to construct a railroad from York to a junction with the Cumberland Valley Railroad at some point between Mechanicsburg and West of the Susquehanna river. In 1857 the "Susquehanna Railroad Company" was chartered to build a railroad from some point on the line of the York and Cumberland Railroad to Sunbury, and the correct geographical alliance of this enterprise with those of the railroad companies previously mentioned was so fully appreciated that they were allowed to subscribe to the capital stock of the Susquehanna Railroad Company, and the city of Baltimore loaned its credit to the extent of $500,000. In 1854 the Legislature of the State of Maryland, and in 1855 that of the Commonwealth of Pennsylvania, passed an act authorizing a consolidation of all the interests between Baltimore and Sunbury, conditioned that there should be no discrimination in favor of Baltimore as against Philadelphia. Provision was also made at the same time for consolidating the amount due by the railway company to the State of Maryland, by the annual payment of $90,000, equal to six per cent. on $1,500,000. Under these auspices, the road was completed to Sunbury prior to the outbreak of the war in 1861, but the enterprise was not remunerative until the fortunes of war concentrated on to the Northern route, via Harrisburg, a large amount of traffic which, under other circumstances, would have been tributary to the Southern line. In February, 1863, the Shamokin Valley and Pottsville Railroad, extending from Sunbury to Mt. Carmel, and opening up an extensive coal region, was leased for nine hundred and ninety-nine years, and in May of the same year the Elmira and Williamsport Railroad was leased for a similar period. In 1866 the unexpired term of lease, held by the Erie Railway Company, of the Elmira, Jefferson and Canandaigua and Chemung Railroads, was assumed by the Northern Central Railway Company, and through effecting these arrangements, control and independent management was obtained of a through line between Baltimore and Canandaigua three hundred and twenty-five miles in length—(the forty miles between Sunbury and Williamsport being practically in the same interest, although owned by the Philadelphia and Erie Railway Company.) A detailed account has been given of the various steps by which the Northern Central Company attained its present position, with the view of informing the reader what an extent of territory the Company covers in its operations. Traffic for Baltimore originating on any of the lines owned directly by the Pennsylvania Railroad, or under its control, comes via Harrisburg. This is especially applicable to business originating in Pittsburg, of which, until the Connellsville route was opened, the Northern Central Railway had a practical monopoly. All the petroleum conveyed to Baltimore, other than that which is procured from West Virginia, is sent via Harrisburg, and all the

trade for Baltimore originating in Western New York, Buffalo
and the Dominion of Canada, pays tribute to the same route. An
impetus could doubtless be given to this traffic if the construction of appropriate terminal facilities at tide-water was carried out.
The necessity for having such conveniences for shipment of produce
has long been realized by the managers, and during 1873 a lease
was entered into with the Canton Company for 700 feet of water
front on the Susquehanna wharf, with a depth of over 1,200 feet
from Third street to the Port Warden's line, and also a large lot
below Eighth avenue, between Thirteenth and Fourteenth streets,
and binding on Twelfth avenue about 1,000 feet, and running to
the water of the Patapsco river, at an annual rent of $15,000 for
the two lots together. The understanding of this lease was that
the first lot was to be improved for the tide-water terminus of the
Northern Central Railway and Baltimore and Potomac Railroad
for the grain, produce and general merchandize trade, and the
other lot on the Patapsco river was to be improved with coal piers
and wharves. It was stated at the time of entering into this lease
that "the lessees would expend at any rate $1,000,000 on the
wharves and terminal improvements," but the outlay has not yet
been made, neither has the stipulated coal tonnage been sent over
the Union Railroad to Canton. The railway company still uses
the mule and horse tracks heretofore laid down on the grades of
Monument street, Central avenue and other streets to tidewater at
Fell's Point, the use of said tracks being perpetuated by the per-
mission of the City Council of Baltimore, in violation, as it is
stated, of an express contract made with the Canton Company for
their removal on the reörganization and reconstruction of the
Union Railroad. It may be noted here that a perpetual lease of
the Northern Central Railway to the Pennsylvania Railroad Company has been agitated during the past eighteen months, and
negotiations looking to such lease have been partly interrupted by
the panic of 1873, and partly by a controversy amongst the stockholders in regard to the rate of dividends to be paid under the lease.
This may account for the delay in carrying out the programme for
improving the leased ground at Canton, or it may be that the
Northern Central Railway Company are awaiting further developments with the expectation that superior terminal facilities will

be brought within its reach at a more convenient and available harbor.

(3.) Baltimore and Potomac Railroad.

It seems extraordinary that while the State of Maryland and the city of Baltimore were devoting their energies to building and perfecting connections with the West and other sections of the country, they should have ignored, to a very great extent, traffic originating within forty or fifty miles of the city in some of the richest agricultural counties of the State, and for which steamers or sailing craft furnished the only means of transportation. The Western Shore was until recently entirely without railroad facilities, and although a charter had been obtained in 1853 for constructing the Baltimore and Potomac Railroad, still no county or State aid could at that time be procured, from the fact that the inhabitants and property owners of that section were willing to "let well alone," and could not be brought to appreciate the stimulus which railroad construction would give to their local industries, and the enhanced value of their property. It may have been that the land owners in these counties were all wealthy and hated any thing like innovation; be that as it may, just as two sections of the road from Marlboro' to Odenton (junction of the Annapolis and Elk Ridge Railroad) had been placed under contract, the war broke out; a period of inactivity followed; the whole formerly existing system of labor was disorganized, and the property owners of the various counties through which the road was located, then alive to the paramount importance of a railroad, were so financially crippled that they could not contribute to its construction. The building also of a competing line was also strenuously opposed by the managers of the Baltimore and Ohio Railroad Company, and it was with great difficulty that the necessary additional legislation for perfecting the work was obtained. This legislation was carried in the session of 1867, and immediately thereafter negotiations were entered into with the Pennsylvania Railroad Company for aid in constructing the road, and as independent control of a road to the National capital, and thence to a connection with the Southern system of railroads, was deemed advisable, the necessary

financial aid was granted, and the road from Washington to Baltimore was opened for traffic in July, 1872, and from Bowie (26
miles from Baltimore) to Pope's Creek, 49 miles, in January, 1873.
The cost of construction was heavy, in consequence of the high
price of labor and material, and the long tunnels under the cities
of Wasington and Baltimore, but the advantages to be derived by
the city of Baltimore from its construction cannot be over-estimated.
New industries will be developed along the new line of road, an
impetus will be given to the production of tobacco, fruit, early
vegetables and cereals, all of which will converge to Baltimore as
a market, while in addition to the development of purely local
resources, the interchange of through traffic with Richmond and
other sections of the South tapped by the connections at Alexandria
and Quantico must increase in a ratio corresponding with the facilities furnished for transportation.   There is no reason why, with a
complete through line from Richmond to Weldon, Wilmington,
Columbia, Augusta, Charleston and Savannah, via the "Atlantic
Coast Line," and with another through line from Richmond to
Charlotte, Danville, Spartanburg, Atlanta, West Point, Mont-
gomery, New Orleans and Mobile, via the Piedmont Air Line,
and with the through traffic of both these well-known through lines
converging to the Baltimore and Potomac Railroad and its connections, Baltimore, as a commercial centre, should not be largely
benefitted.   There is no reason why, under proper management
and with appropriate inducements, (that is, if the South, freed from
its political imbroglios, returns to its normal condition of affluence
and prosperity,) Baltimore should not avail itself of advantages
within its grasp and become in the future, as in the past, the general emporium or market for the merchants of the South.

(4.) Western Maryland Railroad.

While the capitalists and merchants of Baltimore have been
busily occupied in perfecting their railroad connections with
the West, North and South, they have not been unmindful of
the traffic which must result from the construction of a railroad to develop the mineral wealth of Carroll, Frederick and

Washington counties; said railroad eventually, in all probability to be extended to Johnstown, Pa., and to the oil regions. It is believed that the Western Maryland Railroad Company is successor to the franchises of the "Baltimore and Westminster Railroad Company," a company originally formed to purchase from the "Baltimore and Susquehanna Railroad Company," all its rights and interest in a branch road, then being constructed from a point on its main line within ten miles of the city of Baltimore, through Westminster to some point on the Monocacy river in Frederick county; which would form the most convenient and direct practicable communication between that county and the city of Baltimore. The company operating this road had only succeeded in constructing 33 miles from Relay to Union Bridge at the close of the war, and was entirely unable to extend the line to Hagerstown and Williamsport in consequence of financial difficulties. The city authorities of Baltimore, appreciating fully the situation, and realizing that without their aid the objects contemplated in the original charter could not be carried out, came forward at this juncture and loaned the company the credit of the city by endorsement of bonds and issue of city stock. Under these auspices an independent line of road has been built from Fulton Station in Baltimore to Owings' Mills; the line has also been completed from Union Bridge to Williamsport, where connection is made with the Chesapeake and Ohio Canal, and where, as soon as terminal facilities are provided at tide-water, a large coal traffic will originate. A tide-water terminus at Canton has been contemplated, and arrangements were entered into looking to favorable business relations with the Baltimore and Potomac Tunnel and the Union Railroad; but the schedule of rates up to the present time has been prohibitory except for purely local business, which will naturally bear a somewhat higher charge to avoid transfers. The charges from Fulton Station to tide-water are 39 cents per ton or even more, and the gross charge on coal per ton is consequently higher than by other lines; hence, under present circumstances, unless the Western Maryland Railroad can secure some cheaper access to a shipping point, its coal trade must of necessity be confined to local demands. As pertinent to this subject, a quotation is given from the annual report of the Canton Company for 1874. "The rates

of toll demanded by the Baltimore and Potomac Railroad of the
Western Maryland Road for the passage through her tunnel of
coal and other tonnage, and also over the short link of the Northern Central Railway that connects with the Union Road tracks at
North street, are, in fact, prohibitory on a new trade which of
necessity, must be comparatively small for a year or so; but if
fostered and encouraged by moderate rates as it ought to be, would
increase in time to such an amount as, would enable the Western
Maryland to submit to the charge, under the city ordinance which
operates on a sliding scale in proportion to the amount of business
that passes through the Baltimore and Potomac Tunnel." Other
outlets are available for the Western Maryland Railroad, and the
city of Baltimore will be derelict in duty to itself, if, after having
made such heavy advances in the direct interest of the city, she
does not provide that these advances shall be utilized to the best
advantage, and economical tide-water facilities furnished without
delay in accordance with the increase of traffic. A comparatively
cheap line (as to construction,) can be built along the western
limits of the city to the deep water below Spring Gardens, or to
the deeper location at Curtis' Bay, (now Pennington.) It is not
merely however as a purely local road, bringing the inexhaustible
supplies of the Cumberland Basin to tide-water, together with the
agricultural products of Carroll, Frederick and Washington counties, that the Western Maryland Railroad is to be regarded. Those
who originally projected it, and those who are associated with its
present management, have still higher and more extended aspirations for its future; they know that the road can be extended to
Johnstown, Pennsylvania, a distance of 100 miles, at a cost per
mile not to exceed that of the Baltimore and Ohio Railroad; they
know that by construction of such extension, Pittsburg would be
brought within 258 miles of Baltimore, and that every section of
the road would yield a heavy and annually increasing tonnage in
coal, iron ore, lumber and other agricultural produce; all tributary
to the city of Baltimore, all tending to build up her reputation
and prestige as a manufacturing and commercial centre. It may
be noted that by such a line, if operated conjointly with the Pennsylvania and Pittsburg, Fort Wayne and Chicago Railroads, the
distance between Chicago and Baltimore would be reduced to 727

miles, as against 795 miles via Centreton and Newark, and 801 miles via Pittsburg and Harrisburg. This difference representing 3 hours time for passenger travel, and in a corresponding ratio for freight traffic cannot be ignored, and must be availed of by competing lines, especially if our prognostications relative to the commercial prominence of Baltimore are correct, and substantiated by facts.

(5.) *Philadelphia, Wilmington and Baltimore Railroad.*

The existing connections of Baltimore by railroad West, North and South have now been fully alluded to, but it has also a direct Eastern connection with Philadelphia, New York and Boston, over the line of the Philadelphia, Wilmington and Baltimore Railroad. It may be presumed that almost all the traffic passing over this road, from Baltimore to Philadelphia and beyond, is through to and from the South and West; a certain proportion also of the local business, more especially that originating west of Perryville, will doubtless radiate to Baltimore as a market, while east of that point the trade and business travel will go to Philadelphia. The all-rail connection, however eastward, is evidently highly appreciated, and as unusual facilities are furnished by the railroad company for excursion travel, trains during the summer months are patronized to an extent which compensates for the decrease in through business, incident to the recess of Congress. Cars are transferred, without breaking bulk, from the Philadelphia, Wilmington and Baltimore Railroad, in Canton, to Locust Point, and there made up into trains for the West and South. Traffic for the Baltimore and Potomac Railroad connects via the Union Railroad and Tunnel.

(6.) *Baltimore and Drum Point Railroad.*

Prominent among the prospective railroad connections of the city of Baltimore, may be mentioned the Baltimore and Drum Point Railroad, now in process of construction from Baltimore to Drum Point, on the Patuxent River, a distance of about seventy-

four miles. The enterprise has created a great deal of attention
throughout the State, and is regarded favorably by the State
authorities, in view of the fact that Drum Point Harbor is pronounced by officers of the United States Coast Survey to be infe-
rior only to that of Portsmouth, New Hampshire, and has been
for a long time regarded by intelligent merchants and shippers as
a point which would prove a valuable adjunct to the commerce of
Baltimore, if connected with that city by a railroad; affording as
it does the deepest water—never liable to obstruction by ice or
otherwise—and within an easy run of the Capes. Time will of
course demonstrate whether these ideas about establishing a coal
depot at the southern terminus of the new road will be successfully realized, and whether a port, which has no opportunities of
disposing of or distributing inward cargoes, can be made a financial success; but pending the solution of this question it may be
stated that the development of the fine country between Baltimore
and Drum Point, by the construction of a railroad, must inure to
the prosperity both of the State and city, by stimulating the production of the earliest fine fruits and vegetables, which will find a
ready market in Baltimore, both for immediate consumption and
for canning. A large oyster trade can be also built up, and the
contributions of tobacco and grain from Anne Arundel and Cal-
vert counties will be very considerable; as a proof of this it may be
stated that the crop of Maryland tobacco inspected in Baltimore
amounted, in 1872, to 30,000,000 pounds, about one-third of
which was produced in the counties above mentioned. There are
also considerable products of butter, milk, eggs, poultry, meats,
and wood and timber of all kinds. For these productions the
only means of communication with a market have hitherto been
by steamboats, which are frequently debarred from running regu-
larly at a season of the year when their services are most required.
It is estimated by competent judges, that with certain definite
means of transportation guaranteed by a railroad, the local productions of the country could be increased at a very low average
nearly 500 per cent. Another important advantage which Baltimore will derive from the construction of the Baltimore and Drum
Point Railroad, will be the direct route established by it between
the city and the State capital. The distance between Annapolis

and Baltimore by the present route of the Washington Branch and Annapolis and Elk Ridge Railroads is forty miles; by the Baltimore and Drum Point Railroad the distance will be only twenty miles, and will be run within one hour, without change of cars, and at a much less charge. As an evidence of the amount of traffic now carried on between Annapolis and Baltimore, with very imperfect and unsatisfactory arrangements, it may be stated that the revenue derived therefrom exceeds $90,000 per annum. It is difficult to predict what increase may be anticipated when the new road is built and superior conveniences furnished for the traveling and shipping public.

(V.) PRESENT TERMINAL FACILITIES.

In alluding to the existing terminal facilities at Baltimore for transacting a large commerce, coastwise and with European ports, it will be distinctly understood by the reader that reference will only be made to the tide-water termini, and not to those points in the city proper which are only used in connection with local traffic, and to secure its prompt and economical handling. It will be advisable also to ignore, for practical purposes, the harbor and basin of Baltimore, as they are termed, because as shown above in Article III—on the Situation of the Harbor—they are only adapted for small coasting vessels, drawing a few feet of water, and are, under existing circumstances, a serious drawback instead of an advantage to the city. The terminal facilities, therefore, may be considered as comprised in those at Locust Point and Canton.

(1.) Locust Point.

The advantages of Locust Point as a tide-water terminus were appreciated by the managers of the Baltimore and Ohio Railroad at an early period in its history, and in 1848 Mr. Thomas Swann, then President of the Company, purchased some land, and tracks were built for the coal traffic, which increased more than 150 per cent. in the course of two years through what were then considered favorable arrangements for its shipment. In 1851, three years after the first purchase by Mr. Swann, alluded to above, it was found that the lands of the company were entirely inadequate to the requirements of the augmented traffic, and an inducement was held out to private parties to erect their own wharves on the northern front of Whetstone Point. This inducement, which was in the shape of a drawback of six cents per ton on all coal received by parties at their own wharves, has produced very favorable results, as is attested by the numerous wharves now in existence, which are taxed to their full capacity. The arrangement of this system of private wharves at Locust Point is very perfect, and the shutes are constructed so that coal is loaded from the cars directly into the hold of the vessel at a merely nominal cost, the drawback alone guaranteeing a handsome interest on the money invested in the original erection of the wharves. In the twelve years immediately subsequent to the purchase of their property at Locust Point, the managers of the Baltimore and Ohio Railroad Company did little, if any thing, towards its general improvement, otherwise than by securing facilities for handling the coal traffic—but in 1860, after their connections with the West had been established, fresh attention was directed to the advantageous situation of the property as a tide-water terminus, and a European line of steamships was projected. All efforts in this direction were, however, temporarily neutralized by the outbreak of the war, and the discouragement of all public enterprises incident thereto, neither was the subject publicly agitated again until 1865, when, as before stated, the Baltimore and Ohio Railroad Company purchased from the United States Government four steamships, which were named respectively the Alleghany, Carroll, Somerset and Worcester, but which were shortly found to have too limited a carrying capacity for the increasing business, and were consequently superceded by the magnificent Clyde built steamers of the North German Lloyds' Line. The establishment of this steamship line necessitated the immediate construction of steamship piers and warehouses. The piers, wharves and warehouses have been built in the most substantial and approved manner, with every possible precaution against damage by fire or other casualty, and one of the piers is 760 feet long and 90 feet wide, while the other is 675 feet in length with a width of 100 feet. These two piers are covered over with iron sheds, and on each there is a double track, on which the cars are run in directly from the yard. There is a space of 100 feet between each of these two piers, and another space of 100 feet between the second steamship pier and the next wharf. It was estimated in the construction of these docks that facilities would be furnished for loading and unloading, if necessary, at one time four large steamships. Freight can be transferred directly from the hold of the steamer to the cars, and vice versâ. In addition there is an extensive bonded warehouse on the pier, and in all the improvements, as would be evident to any careful and intelligent observer, the economical and expeditious handling of merchandise has been kept constantly in view. One of these piers is now devoted to the through traffic to or from New York, Philadelphia and Boston, destined for the main line, which is brought by the various steamers running to these Eastern cities, the volume of which is much larger than would be anticipated from the strong competition existing for transportation between the East and West. Apropos of these docks, it may be noted that the visitor is very forcibly struck with their admirable system of construction and general arrangement. All the buildings along the water-front are surrounded on three sides by water, the docks are, with but one exception, one hundred feet wide, and they are presumed to be of an uniform depth of twenty-four feet; although it is believed that this depth is maintained by continual dredging, there being no active current to carry away accumulating deposits. There are at present, seven docks; the first at the western limit of the company's water front, between the two steamship piers above mentioned; the second, between the second steamship pier and an extensive wharf; the third, between the wharf and the small elevator; the fourth, between the small elevator and the railroad ferry; the fifth, between the railroad ferry and what soon will be the coffee warehouse; the sixth, between the proposed coffee warehouse and the large new elevator; the seventh, between the last mentioned elevator and a wharf which runs back to the walls of the Fort. The improvements in connection with the first, second, third, fourth and sixth docks may be considered as complete; between the fifth and sixth docks, the company are engaged in constructing after the most substantial and approved designs, a large coffee warehouse. This warehouse will be more than 267 feet long by 77 feet wide. It will consist of two stories and will be absolutely fire-proof. The massive foundation of this warehouse rests on piles, or rather on a substantial frame-work placed on piles and braced securely, with a view of obviating any ultimate settling of the foundation. Stone, brick and iron are the only materials which will enter into the construction of this warehouse, and single tracks capable of accommodating nine cars each, will be placed on each side. A similar warehouse with equal facilities for handling traffic, will be built on the wharf contiguous to the seventh dock, and this will be devoted to the storage of sugar. The main object of the company in erecting these two substantial warehouses for sugar and coffee, is to foster the rapidly increasing trade between Baltimore and the West Indies or South America, and to stimulate the importation of these two staple articles for the West and North-west via Baltimore; in fact, when all these plans are consummated, Baltimore will become the most convenient and economical port of entry for commodities which have heretofore been distributed from New York. The commercial interests of the city will also be much benefited by these improvements, through the attraction of a much larger number of vessels to Baltimore as a port of entry; hitherto, conveniences have been furnished for a large export trade, and but, comparatively speaking, little attention has been paid to the storage requirements of an import trade; hereafter, vessels plying between Baltimore and foreign ports can rely on a full cargo both ways, the rates of freight can be correspondingly reduced, and the consumers of the West equally with its producers, will derive substantial advantages from a reduction in first cost, and from a low rate both on ocean and inland transportation. In connection with its system of wharves, docks and warehouses, the Baltimore and Ohio Railroad Company has established at Locust Point a railroad ferry across to Canton, where it connects with the Philadelphia, Wilmington and Baltimore Railroad. There are two slips, each 40 feet wide, into which barges are run having a capacity of ten cars each; and it is stated that during the busy season, as many as 250 cars a day are transferred; in fact, by having such facilities, the company obviates entirely a tedious and expensive transfer through the city, and a large through business is transacted between New York and Philadelphia, or the South and West via the Baltimore and Ohio Railroad. The visitor however, at Locust Point is perhaps struck most forcibly with the elevators there contiguous to the fourth and sixth docks. Both of these elevators are substantially built after the most approved design, and have conjointly a capacity of 2,100,000 bushels. The arrangements for handling grain elicit the unqualified admiration of all who have

5

inspected them and are deservedly pronounced superior to any now
in use on the Atlantic Sea-board. The smaller elevator, (the first
built,) is 150 feet in length by 80 feet in width, with 120 bins, 9
feet 6 inches square and 65 feet in depth, and eight elevators, five
for receiving and three for shipping grain. The large elevator is
324 feet 10 inches long, 96 feet 10 inches wide and 168 feet 10
inches high, it is worked by two engines of 400 horse power, and
has 16 receiving with 8 shipping elevators. The bins are 210 in
number, 11 feet six inches square on the inside and 94 feet 3 inches
deep. There are 300 buckets attached to the belt of each elevator,
and their capacity is estimated at 100 bushels per minute, in fact the
facilities for handling grain are almost unlimited. The large elevator is approached by four sets of tracks, two of which run directly
into the elevator building. The reader will easily comprehend
that in utilizing to the fullest extent the 80 acres of land, which
they own at Locust Point, the engineer of the Baltimore and Ohio
Railroad Company has exercised consummate skill : the coal traffic
has its special branches, with sidings and other accommodations ;
the steamship piers have their own peculiar tracks, switches and
turn-outs, while for what is known as the New York Line, and in
connection with the railroad ferry, a new road has been built along
the southern water front of the peninsula. In this labyrinth of
tracks, duly connected with each other by switches, so as to secure
in all instances quick handling of the cars, are more than twenty
miles of iron and steel, and this track mileage will be further increased when the company's land between the new elevator and the
walls of Fort McHenry is extended and improved. And yet with
all these facilities, with a capacity for handling one thousand cars
daily, exclusive of the coal traffic, a question arises whether with
their limited area the Company will be able to accommodate in its
present location the large accretions of traffic, which must naturally
result from the opening of a direct avenue to Chicago and the
North-West. The crowded condition of the piers indicates very
clearly that they are now taxed to the utmost capacity : no further
extension of water front can be secured, unless by the purchase of
the Fort McHenry property from the government, which can
hardly be anticipated, and it would be wise for the railroad company, even in this comparative infancy of its export and import

trade to make definite arrangements for securing another tide water terminus, to which a large proportion of its coal and petroleum business could be transferred, and the yard or grounds at Locust Point reserved exclusively for grain, coffee, sugar, and other general merchandise; also for its through passenger traffic which will it is believed, be shortly diverted from Camden Station to this point, and be thence transferred by ferry to a connection with the Philadelphia, Wilmington and Baltimore Railroad at Canton. Much has been done already, and that much in the most perfect systematic and substantial manner; but this much will almost dwindle into insignificance within ten years from this time; i. e. if the increase of traffic during the next eight years, is commensurate with that which has characterized a similar period in the past. There is no reason why the coal trade from Cumberland should not increase from 2,000,000 to 5,000,000 tons a year with convenient outlet at tide-water; business of any character responds to the facilities furnished, and the efforts of the Philadelphia and Reading Railroad Company, in the direction of building up a mammoth coal trade, together with the uniform success which has attended such efforts, should convince all intelligent business men that what is now a remunerative trade might be doubled, nay, quadrupled, by a correct realization of the future and an anticipation of its numerous wants.

(2.) Canton.

The second existing tide-water terminus of the city of Baltimore, is at Canton, on the east side, and the Company owning the property, about 2,400 acres, with 32,000 feet, or more than six miles of water front, was incorporated in 1828. Extensive privileges were conferred by the Charter or Act of Incorporation, as will be seen from the following extract:

"The objects for which the Canton Company of Baltimore aforesaid are incorporated, and which the said Company are hereby authorized to effect, are the improvements in such manner as shall be conformable to the laws of the State, and not contrary to nor inconsistent with any of the rights and privileges of the corpora-

tion of the city of Baltimore, or of any citizen or citizens of this
State or of the United States, of any lands and appurtenances
which shall belong to said Company, by laying out into lots,
streets, squares, lanes, alleys, and other divisions, any such lands
within the vicinity of the city of Baltimore, or near to any navigable water, and erecting, constructing and making thereon 'all such
wharves, ships, boats and other vessels, workshops, factories, warehouses, stores, dwellings, and such other buildings and improvements as may be found or deemed necessary, ornamental or convenient; and letting, selling, leasing, renting or granting on con-
ditions, or using any lot or any other portion of the said lands for
agricultural, mining or manufacturing purposes; or any wharf,
house, or other building or improvements to be used by any
mechanic or artizan, or other person, whether in the employ of
said Company or not, in carrying on any lawful trade, business or
manufacture authorized or permitted by the laws of the State."
When the Canton Company was first incorporated with its capital
stock of $2,000,000, sub-divided into 20,000 shares of $100 each,
the brightest future was confidently predicted for its operations.
The stock became a foot-ball, so to speak, of speculators and
adventurers, at one time, on the strength of an alliance with the
Northern Central Railway, and the construction of its line to tide-
water on the Company's property; at another, of a probable connection with the Western Maryland Railroad. The fluctuations in
the price of the stock were prior to the panic of 1857, almost in-
credible, ranging from $54 per $100 share to $260; but in the
three years immediately succeeding 1857, the stock passed into the
hands of some enterprising gentlemen who appreciated its value,
and addressed themselves unhesitatingly to carrying out the objects
for which the Company had been originally incorporated; viz., a
substantial improvement of the land by leases on ground rent, and
by the construction of wharves. The proceedings of the Canton
Company, however, were paralyzed by the same commercial inactivity and depression—which depressed the whole State of
Maryland, during the war from 1861 to 1865. In 1866, a company, called the "Union Railroad Company," was incorporated
with power to construct a railroad from Relay, on the Northern
Central Railway, where a connection was also made with the
Western Maryland Railroad, to Canton, by a route which would

carry the road around the city, passing down to Canton from the north. The capital stock of the Union Railroad Company was fixed at $600,000, and authority was granted to commence work as soon as $150,000 of the share capital had been actually taken up. Of such $150,000, one-third was subscribed by the Western Maryland Railroad, and one-third by the Canton Company, the balance by private individuals. Such a road would practically have constituted an extension of the Western Maryland Railroad and would not have answered the purpose of accommodating the city, neither would it have developed the Canton property in accordance with the original design of its stockholders and proprietors. Another plan was subsequently suggested, looking to a connection with the Western Maryland Railroad at or near Owing's Mills and with the Northern Central Railway at or near the city. In furtherance of this plan the city authorities agreed to endorse the bonds of the railroad to the extent $500,000, and aid was also promised by the Northern Central Railway Company, but experience soon demonstrated the utter folly of attempting to carry out such a heavy undertaking with a limited capital, and it was not surprising that after having secured the endorsement of the city to $117,000 of bonds and having expended of such sum $100,000, the work was suspended and a lamentable failure recorded. Again in 1870, through the efforts of the Canton Company additional legislation was obtained for the Union Railroad Company, and it was proposed that each of the railroads centering in Baltimore should join hands with the Canton Company and build a road to tide water, it being understood that the tolls to be levied on freight passing over such road should be fixed by a committee representing all the parties interested. The scheme appeared to all intents and purposes feasible, but its originators were unaware of the rivalry existing between conflicting interests, and that it was almost impossible to harmonize difficulties even when such amicable settlement would have tended equally to the advantage of the railroad companies, and to the commercial prosperity of the whole city. The so-called federation scheme failed, but in this emergency, when all hopes of resuscitating the enterprise were apparently lost, the Canton Company, appreciating fully the advantages which would accrue to their property from the completion of a railroad to their tide-water privileges, unanimously determined to subscribe

for the whole capital stock of the Union Railroad Company, and
if necessary, endorse its bonds. Under these auspices, after new
surveys had been made and new estimates prepared, the work was
commenced in 1871 and completed in June, 1873, at a cost, including the tunnel, of $2,300,000. The Union Railroad com-
mences at a junction with the track of the Northern Central
Railway, near Charles street, and its terminus is at tide-water at
Canton; and, together with its eastward branch, connecting with
the Philadelphia, Wilmington and Baltimore Railroad, makes
about seven and a half miles of double track, or, with sidings and
switches, about eighteen miles of rail. It is laid with a heavy 72
lb. iron rail. The tunnel is five-eighths of a mile long, and is
double arched. About three miles from the junction with the
Northern Central Railway near Charles street, the road bifurcates,
one branch connecting with the Philadelphia, Wilmington and
Baltimore Railroad, and establishing an all rail through route to
Philadelphia and New York; the other track goes due south,
passing through the lands of the Canton Company for three
miles, until it reaches the terminus at tide-water on Ninth street.
The land required by the Union Railroad Company, including
right of way for double track with ample facilities for water stations, turn tables and other terminal improvements, amounted to
thirty-five acres, and this has been deeded to the railroad company
at a consideration of $200,000. The railroad, although managed
by an independent organization, is practically the property of the
Canton Company, as they own 5,940 out of 6,000 shares of
common stock and have endorsed its bonds to the amount of
$1,383,000—$783,000 being six per cent. currency, first mortgage,
and $600,000 seven per cent. gold bonds. There is also an amount
due the Canton Company for land, $200,000. In making these
heavy outlays for completing the Union Railroad, amounting in the
aggregate (as stated above and including $117,000 first mortgage
six per cent. bonds endorsed by the city) to $2,300,000, the Canton
Company estimated that in addition to making a valuable permanent improvement to their property, they were securing a highly
remunerative investment. It was estimated that at any rate
300,000 passengers would travel over the road at thirty-five cents
each, also that 2,000,000 tons of coal, iron, lumber, grain, oil, and
other merchandise would be transported the whole distance of

seven miles at a fixed tariff of five cents per ton per mile. The total direct revenues from this investment would have been, according to the estimate, about $805,000, and after paying the charges for maintenance there would have been an ample surplus to pay the annual interest on the bonded debt and a handsome dividend on the share capital. Unfortunately these sanguine expectations have not been realized, but from causes entirely beyond the control of the directors of the Canton Company or the managers of the Union Railroad, and in accordance with a contract entered into with the Northern Central Railway Company, the Baltimore and Potomac Railroad Company, and the Western Maryland Railroad Company, for a period of 99 years from December 30th, 1873, a tariff for freight and passengers has been fixed, which accords to the Union Railroad twenty cents per ton and fifteen cents per passenger to or from tide-water at Canton and to or from the Bayview Junction of the Philadelphia, Wilmington and Baltimore Railroad. A considerable revenue, even at this decreased rate of charges, could be made, if the traffic were concentrated on the road as had been originally anticipated. A promise made by the President of the Northern Central Railway, that one million tons of coal should be sent over the Union Railroad within twelve months after its completion to tide-water, has entirely fallen through ; the Baltimore and Potomac Railroad has entirely failed to furnish its quota of merchandise and general traffic, while little, if any, revenue has been derived from the Western Maryland Railroad, through the prohibitory toll demanded by the Baltimore and Potomac Railroad Company on all business passing through its tunnel. There is no doubt that the Canton property is in some important respects very valuable, and its value would be very much enhanced if all the railroads converging to Baltimore came in on the east, and not the west or north side of the city. The only railroad whose natural geographical tide-water terminus is at Canton, is the Philadelphia, Wilmington and Baltimore Railroad ; and it is believed that arrangements will shortly be made for transferring the President Street Depot to a lot on the Canton property having 177 feet water front, where extensive improvements in the shape of wharves and piers are contemplated ; but in anticipation of the opening of the Union Railroad and the concentration on it of the traffic of several railroads mentioned above,

many large factories and extensive improvements have been projected. A large grain elevator will be erected to meet the requirements of the grain trade. This will be in addition to a transfer
elevator of 100,000 bushels capacity at what is known as "Gardner's Union Railroad Depot and Elevator," which, according to
the latest reports published, is transacting a heavy business—the
receipts for five months ending May 31st, 1874, having been
5,277,000 bushels, a fraction less than 2,000,000 short of the
entire year 1873, or 2,327,000 bushels in excess of the corresponding months of 1873. An extensive car wheel works is also in
operation, and locomotive, car and carriage building works are also
projected. Oyster and fruit packing houses, foundries of various
sorts, blast furnaces, fertilizer manufacturers, sugar refineries, steam
saw mills, planing mills and sash factories, distilleries, coal oil
refineries, breweries, chair and furniture factories, together with
other manufacturing establishments, are already built and in operation on the grounds of the company. A large and thrifty larboring
population is being attracted to Canton as a residence, and even if
the railroad companies fail to make Canton their principal tide-
water terminus, still with its existing facilities for manufactories
it will enable the company to dispose of their lands at very
remunerative figures. By reference to the map it will be seen
that the water front of the Canton property is extensive, extending from Washington street, at the base of Fell's Point to the
Lazaretto Point, forming the western boundary of the North-west
Branch, lying opposite Fell's Point and Locust Point, making a
water front of more than five thousand feet, with a depth of water
varying from sixteen to twenty-six feet deep at the Port Warden's
line. From Lazaretto Point, extending eastward, into Colgate's
Creek, there are fifteen thousand feet water front. These fronts
are in straight lines, and the number of feet could be greatly increased by the erection of piers and docks. Out of this water
front the Canton Company still holds among its assets, 18,750 feet
and in addition, 18,500 building lots and 900 acres of land. In
concluding this brief review of the terminal facilities at Canton, it
might be appropriate to quote the remarks of the managers of the
Company in estimating the value of their property, in comparison
with that at Port Richmond, Philadelphia: "only a few years
ago that now most important quarter of Philadelphia was a thrift-

less long shore village. The Reading Railroad waked it into life. More than 20,000 vessels received 2,800,000 tons of coal, at its one-and-a-half miles of double piered wharves in one year. A city has supplanted gardens, and land has risen from $25 to $25,000 per acre. In contrast with Port Richmond, the advantages are largely on the side of Canton. The Port Richmond Railroad is but a local coal bearing road. The Union is its equal as a coal bearing road, and the outlet of a system of roads which penetrates three-fourths of the States in the Union. Richmond is above Philadelphia, and has no advantages of water. Canton is below Baltimore, and owns four-fifths of all the deep water on the north side of the harbor. The area at Canton is ample as that at Richmond. At its wharves, ships of the greatest burthen may receive their freight direct from cars. On its spacious lands, there is room for depots convenient to store on a level with the cars the freight too heavy to hoist—for furnaces, forges, factories, mills and shops, into which trucks may carry material too heavy to bear the cost of frequent handling, in competition with cheaper labor; for the combination of land and water necessary to build and launch vessels, whether of wood or iron, where materials may be raised from the train to the stocks; and on the high hills, sites for factories, where the breeze will cool the air, that women and children breathe as they tread their spindles and the loom, and whence in return the trains will convey direct from factory, shop and ship, the products of Eastern labor to consumers in the far-off West, no longer enhanced in prices by storage, drayage, or commissions."

It has been our aim in presenting our readers with an account of the terminal facilities of Baltimore, as they now exist, to present a succinct and truthful statement. The facts, as given, are partly the result of personal examination, and are compiled in part from documents published under the auspices of the Baltimore and Ohio and Canton Companies. The exertions made to develop these terminal facilities and build up the trade and commerce of Baltimore should convince all that the owners are really in earnest, and have implicit confidence in the bright future of their city. They act with a faith that the future of Baltimore is whatever its citizens may choose to make it, and that natural advantages, superior to those of any other seaport on the Atlantic seaboard are supplemented by new and varied elements of prosperity and development.

(VI.) DIFFICULTIES IN THE WAY OF FURTHER DEVELOPMENT AT EXISTING TERMINI.

Allusion has now been made at some length and it is hoped truthfully and impartially to the early foundation, growth and present condition of Baltimore, to its advantages of location for commercial and manufacturing purposes, to the situation of its harbor, its railroad connections and its present terminal facilities, it will now be demonstrated that there are difficulties in the way of further development at existing termini. And first as regards *Locust Point* the present tide water terminus of the Baltimore and Ohio Railroad. It is true that the company has utilized to the fullest extent the area of eighty acres at its disposal, it is true that the labyrinth of tracks systematically and economically arranged, bears abundant evidence equally with the magnificent warehouses and elevators and the substantial piers to the consistency with which the managers of the road have pursued the idea of placing the commerce of Baltimore on a firm and it might be truly said immovable basis, but the traffic has increased so rapidly since these improvements were first inaugurated, that even now from all appearances the steamship piers are taxed to their utmost capacity and a much larger water frontage would if practicable be secured to meet the requirements of an increasing business. The question arises can it be secured? and even if secured can it be rendered available for the purposes contemplated without an expenditure of labor, material and money hardly in accordance with the requirements? The site now occupied by private coal wharves might be purchased and warehouses for storage or commission merchants erected thereon, piers and wharves might be built at some isolated points along the southern front of the present terminus, but the extension of such facilities would only enable the company to provide for the grain and other merchandise traffic, and the coal, lumber and petroleum traffic must be sent to some other port equally convenient as far as depth of water and rail communication are concerned. The policy of concentrating at one point, petroleum, coal, lumber, general merchandise and grain is justly liable to criticism, the risk of accident and loss by fire, in the face of every precaution to the contrary, is intensified, and a railroad company is culpable for hazarding the success of its future operations by neglecting to discriminate and separate its various classes of traffic. It must be remembered, that the prospective business of the Baltimore and Ohio Railroad and its western connections, is destined to be much more diversified than formerly, in consequence of the much larger area of country which will be rendered tributary to it; each special item of traffic has its idiosyncracy, so to speak, and the volume of each will increase in a ratio corresponding with the facilities furnished for its storage, if necessary, and for its prompt and economical handling. It is not merely enough that there shall be unrivaled elevators for the grain trade, or substantial warehouses for the coffee and sugar; provision has also to be made for the cotton crop of the South, for the tobacco of Ohio and Kentucky, for the wool of Illinois and other Western states, for cheese, bacon, lard, packed meats and miscellaneous articles, which now enter very largely into our present European export trade. Traffic will converge to the point where the greatest facilities are furnished, and heavy outlays will be necessary to perfect the plan which has been so auspiciously and systematically commenced. The secret of success in dealing with the public, is to anticipate their wants; and if this rule holds good generally, how much more applicable is it to a railroad corporation, whose prosperity hinges on a correct appreciation of the public requirements and an intelligent understanding of the future. It may be, that these anticipations as to the prospective requirements and increased traffic of the Baltimore and Ohio Railroad are sanguine, and premature; it may be, that the growth of the import and export trade will not be as rapid as has been supposed from the comparative returns of the past three years; it may be, that the existing facilities at the terminus will not be severely taxed by the business originating on 268 miles of new road, all of which will concentrate at Baltimore; it may be, that the coal oil trade will remain stationery, and that the coal and lumber transportation will be identical in volume with that of former years;—

all these contingencies are possible—but, are they probable? Is it not more resonable to anticipate, that with the country recovering from the abnormal condition of depression under which it has labored during the past year, as a result of the panic; with the South especially, restored to affluence and prosperity, business will be transacted on a larger scale than ever heretofore, and a fresh impetus will be given, both to European trade and to all other branches of industry? The careful student of such matters, arguing from analogy, and with a knowledge of the indomitable energy and perseverence of the American people will, it is thought, candidly admit that anticipations relative to the accommodations which will be required at Locust Point within the next two years, are not unfounded; and that, in providing for the grain and general merchandise traffic, apart from the coal, lumber and petroleum, the existing terminal facilities will be taxed to the utmost. The question naturally arises, how and where can the heavy traffic in the three last named articles be accommodated; without increasing the cost of transportation beyond what it is at present? But, before answering this pertinent question, it may be assumed that it is impolitic to concentrate coal, lumber and petroleum in the same yard with general merchandise, or even in such close proximity to a large city, as Locust Point. An outlet for coal, lumber and petroleum can be found at Curtis' Bay, by a line which will save at least five miles of rail transportation, and where, at a moderate expense, wharves and other facilities can be furnished close to natural deep water, extending along two and a half miles of water front. If the owners of coal wharves at Locust Point can dispose of their property there at a very remunerative figure and procure equal, if not superior, facilities for shipment at a much reduced cost, is it not reasonable to suppose that they will avail themselves of the opportunity? If the lumber of Michigan and Indiana can be shipped as promptly and economically from Curtis' Bay as from the wharves of the Baltimore and Ohio Railroad at Locust Point, all lumber for export would concentrate there. The same rule would hold good to petroleum, only in a greater degree, because at Curtis' Bay and its vicinity refineries could be erected, and the manufacture of coal oil, naptha, &c., carried on to an unlimited extent. A railroad can be built from Curtis' Bay to connect with the Baltimore and Ohio Railroad Company at or near Relay, a distance of five and a half miles. The transfer of the heavy tonnage in coal, lumber and petroleum to this short line would relieve very materially the main stem of the Baltimore and Ohio Railroad, and would enable its managers to leave the track from Relay to Locust Point or Camden entirely for passengers and for general merchandise.

Again, at Canton, on the east side of the city, where it has been anticipated by the owners of the property and others interested in real estate, that all the railroads other than the Baltimore and Ohio would concentrate their business : there are serious difficulties in the way of further development which cannot be readily ignored. There is no doubt that Canton is a valuable location for manufactories of all descriptions, and the enhanced value of real estate will compensate the Canton Company for any risk which they have undertaken in making the investment and developing the property. The Union Railroad and Tunnel can also be made remunerative, from the tolls collected on through traffic and from the charges on raw material, coal, &c., consumed by the various manufacturing establishments. The wharf property can also be made available for the shipment of goods manufactured at Canton, and doubtless many canning and packing establishments will be erected in proximity to the water front, where coasting craft bringing oysters, fish, and other produce to a market, can return laden with fertilizers. It will also be an available location for iron furnaces and rolling mills, but it is very much doubted whether Canton will ever become such a general shipping point for foreign trade as has been intimated and anticipated by its owners. By reference to the map accompanying this pamphlet, it will be seen that the railroads centering at Baltimore, with one exception, reach the city limits on the north and west; here are their natural and geographical termini, and it is hardly to be anticipated that the managers of these lines will force traffic down to Canton and through the city at a heavy rate of toll, when they can utilize their own existing lines to better advantage and obtain an outlet to tide-water at a much more moderate expense than they are now subjected to. Take the Northern Central Railway for example. Its managers have, it is true, leased, as stated above, some valuable water frontage from the Canton Company, and it was stated some two years ago that they contemplated an immediate expenditure of $1,000,000 to provide terminal facilities in the shape of grain elevators, coal wharves, &c.; but, up to the present time, no such expenditure has been made, and whether the delay is attributable to an uncertainty about leasing their road to the Pennsylvania Railroad Company, or whether they have taken better counsel on the subject, is immaterial to the point at issue; one significant fact is indisputable, viz: that, in the face of city ordinances and contracts, looking to the removal and discontinuance of the old horse and mule tracks down Central avenue to Fell's Point, the tonnage, or rather the greater bulk of it, is carried to tide-water by horse power, and the Union Railroad only receives a small proportion of the business. Had the Northern Central Railway Company a pecuniary interest in the Union Railroad and Canton property, the case might be different, and the traffic to tide-water might be diverted from its peculiar geographical route; but the present interest of the Northern Central Railway Company, is to utilize its own track and the Baltimore and Potomac tunnel as far as practicable, if, by such a course, they can obtain a greater direct revenue for an enterprise in which they are pecuniarily interested, and at the same time procure equal tide-water facilities for an export and import trade. For example; under the existing tariff over the Union Railroad to Canton, a car load of coal or other produce would pay an arbitrary fixed toll of $2,00 per car. All this accrues to an *outside and independent corporation.* If the same car load were taken through the Baltimore and Potomac Tunnel, and thence by a line projected on the western limits of the city to the new tide-water terminus at Curtis' Bay, a large proportion of the tariff would accrue directly to the tunnel, and render that costly property more remunerative; in fact, the Northern Central Railway Company would be instrumental in building up, for an enterprise in which they are deeply interested, a traffic which, if tributary to Canton would result in no pecuniary benefit to themselves. The cost of the Tunnel proper, was $2,500,000, and it may be estimated that the other works from a point where the Union Railroad diverges · to a point west of the Tunnel, where the new projected West side

Railroad will run to Curtis' Bay, have cost $500,000 more; hence this short section of road should be made to earn at any rate $260,000 gross, in order to provide the interest on the cost and the annual repairs. The coal, lumber and petroleum traffic originating on the line of the Northern Central Railway, if carried ' from the junction with the Union Railroad to tide-water at Curtis' Bay, at the same tariff as that fixed by the Union Railroad Company, would at once produce a revenue for the Baltimore and Potomac Railroad Company, amply sufficient to meet the greater proportion of the interest on the cost of construction together with the expenses of maintenance, while in succeeding years, the revenue, after deducting operating expenses, would be in all probability largely in excess of interest requirements. Of course, there may be contracts in existence which prevent the consummation of such a scheme, and the Northern Central Railway Company may be obligated to adhere to agreements which are manifestly at variance with their own present and prospective interests; but if such is not the case, and they are untrammelled, except in so far as a lease of property at Canton is concerned, the proper course to be pursued would appear to be remarkably plain and easy of solution. A certain amount of traffic will naturally go to Canton, to and from the various manufacturing establishments now in existence there; grain may and will, in all probability, find its way to the Gardner elevator at that point; but the bulk of the tonnage should, under ordinary circumstances, pay tribute to the road with which the Northern Central Railway Company is peculiarly identified. Again, it seems unnatural that traffic, originating on the line of the Baltimore and Potomac Railroad and its Southern connections destined for shipment from Baltimore, should be sent fully six miles out of its way to tide-water at a cost which must increase its cost to the consumer very materially. All the Southern trade could be sent to tide-water at Curtis' Bay on the line to be constructed from Relay to the water front, (5½ miles,) where ample storage facilities will be furnished for cotton, tobacco, hemp and all other Southern products. The Western Maryland Railroad is also a completed line of railroad terminating at Baltimore, and it was always imagined that Canton was its objective point, and that its tide-water terminus would be on the east side of the city.

To obtain control of the traffic of that road, and to build up at Canton a large depot for the products of the Cumberland Coal Basin from Williamsport, was the cherished idea of the Union Railroad Company, but this very plan appears to have been unintentionally yet *decisively* defeated by the geographical location of the Western Maryland extension from Owings' Mills to Fulton street, at the west end of the Baltimore and Potomac tunnel, and by the prohibitory tariff between that point and tide-water at Canton. Forty cents per ton is a charge which precludes competition, and efforts are now being made to secure an economical line to tide-water on the west side of the city. Surveys have been made of a line from Curtis' Bay to a connection with the Western Maryland Railroad; a company has been duly organized, under the general law of the State, to prosecute the work, and there is little doubt that in the early spring Cumberland coal can be transported to tide-water from Williamsport at a rate which will enable the Western Maryland Railroad Company to double, if not quadruple, its existing coal business, and to compete on equal terms for coal traffic with the Baltimore and Ohio Railroad Company. In alluding thus at considerable length to the causes which must, in all probability, prevent such a concentration of traffic at Canton as has been anticipated, no mention has been made of the fact that in consequence of the heavy filling and piling out to deep water, the cost of construction of terminal facilities must be largely in excess of what it would be at Curtis' Bay, nor that the whole drift of the wind up Chesapeake bay is against the Canton property, while the wharfage at Locust Point and Curtis' Bay is sheltered and, in the latter case, almost land-locked. The Canton property is valuable as an investment, and its value as a site for manufactories of every description must increase annually as the advantages of Baltimore as a manufacturing centre are duly appreciated; but its importance as a prospective shipping point for an extensive railroad system, has been highly exaggerated; and the difficulties in the way of a further extended development of terminal facilities at that point are, from an impartial stand-point, almost insuperable, and it is believed that experience and time will justify the conclusions now arrived at. The reader will note that no reference has been made in this connection to the difficulties of developing further terminal facilities at the harbor of Baltimore proper—the basin as it is termed. The filling up of this semi-stagnant pool, and obviating thereby a large annual expense now entailed on the city, is merely a question of time. Whether that time will be accelerated by the presence of a malignant epidemic consequent on the accumulation of such noxious deposits, cannot now be safely predicted, but the filling up of the present harbor across from Fell's Point, and the utilization of such newly-made ground for the erection of substantial warehouses, is one which should recommend itself to the city authorities *as a direct source of revenue, not of expense.* It is a project which should be regarded favorably by all who are interested in perpetuating the city's enviable hygienic record.

(VII.) CURTIS' BAY—ITS ADVANTAGES AS A PORT AND MANUFACTURING CENTRE.

CURTIS' BAY is situated about two miles due south from Baltimore, on the west side of Chesapeake Bay, and has an entrance between what are known as Fishing and Leading Points of more than a mile wide, accessible for vessels of any draft of water at *all* times, and for sailing vessels in *every* quarter of the wind, except due west, and even then there is abundant room (there being no shoals or reefs) for beating into the harbor. It appears extraordinary that the owners of this valuable harbor and water front have not attempted at an earlier date to develop the property which they have held for nearly a quarter of a century, and which, as including a water front of more than two and a half miles in addition to nearly 1,200 acres of fine land available for residences and manufactories, must be a prolific source of wealth to the present owners, " The Patapsco Land Company of Baltimore City ; " but it may be presumed that there were good and substantial reasons for such apparent inactivity, and the delay may have been advisable in view of the fact that it is only within the last two years that the railroad system of Baltimore has been perfected by the completion of the Baltimore and Potomac and Western Maryland Railroads, and that development of property in anticipation of the requirements of traffic might have been premature. Now, however, when every thing indicates very clearly a bright commercial future for Baltimore, when its capacities for a large export trade can only be limited by the facilities furnished for shipment, and when by the completion of a new line to Chicago it may be anticipated that the demand for available wharf privileges will be considerably in excess of the supply, it has been determined to bring Curtis' Bay and its superior advantages of location prominently to the notice of railroad managers, manufacturers, shippers and capitalists, and it will be surprising if these representations, truthfully and impartially made, do not attract close and patient investigation ; it will be extraordinary if the most commodious and the safest harbor this side of New York, with the exception of Hampton Roads, (to which it is pronounced equal,) is not fully appreciated. Much has been said and written about the harbor facilities at Locust Point and Canton, but they cannot be compared with those which exist *naturally*, not *artificially*, at Curtis' Bay. Twenty feet can be obtained at Locust Point by persistent dredging and removal of the sedimentary deposits, and the docks of the Baltimore and Ohio Railroad Company are presumed to have a uniform depth of twenty-four feet, and yet, as stated in Article III of this pamphlet, it is not many weeks since a steamer drawing nineteen feet of water was hard and fast aground within a biscuit toss of the large grain elevator of the railroad company. Channels for the coal vessels going to the coal wharves at Locust Point have also to be continually dredged out. At various points also on the Canton property water can be found averaging twenty to twenty-four feet, but these are *exceptional* cases, and in the majority of instances piers have to be built out a long distance into the water, so as to furnish suitable accommodations for vessels of heavy tonnage; but that which is *exceptional* at Canton is *the rule* at Curtis' Bay, and while certain merits are willingly conceded both to the Locust Point and Canton stations, while due merit is conceded to the enterprise and energy with which permanent developments have been carried on by the managers and owners of either property— still "The Patapsco Land Company of Baltimore City" claim that at Curtis' Bay the average depth of water is much greater than either at Locust Point or Canton; they claim that such average depth, according to the United States Coast Survey, is more than twenty-four feet at mean low tide, and that this depth can be found within one hundred and, in most instances, within fifty feet of the main land. Another decided advantage of Curtis' Bay as a shipping port lies in its accessibility without the use of a steam tug. Vessels coming up Chesapeake Bay can, when off the entrance to Curtis' Bay, shape their course direct to their wharves; on the other hand, those destined for Locust Point or Canton, are compelled in consequence of the sinuosity of the channel above Fishing Point, to take a steam tug, and the expenses of coming into port are thereby materially increased. By reference to the map, it will be seen

that, according to the surveys, twenty feet of water can be found in
Marley's Creek, fully two miles from the entrance to Curtis' Bay.
It is claimed that Canton and the property of "The Patapsco Land
Company" are equidistant from the business portion of Baltimore
city, while a vessel docking at Curtis' Bay has four miles less to
traverse before reaching its destination, than if it went to Canton.
In addition to these advantages of water location, it will be seen
that Curtis' Bay is much nearer by land to the principal trunk
railroad running to the West and South than Locust Point; the
distance from tide-water to Relay is estimated at five and a half
miles, consequently shipments from the West or South, consigned
to Curtis' Bay, save in distance of rail transportation as against
Locust Point, and this saving would be more especially noticeable
in the coal traffic over the Baltimore and Ohio Railroad, and
in that which may reasonably be expected from the Western
Maryland Railroad.   Allusion has been made in Article VI to
the various difficulties which prevent the further development
of terminal facilities at Locust Point; everything which the experience and ingenuity of man could suggest, has been done to
utilize a contracted strip of land, and make it available for the
shipments of coal, grain, oil, cotton and tobacco; but the fact is
patent that the area of land is not sufficient to meet the requirements of all the traffic offering, and it seems absurd for a railroad
company or individuals to give fabulous prices for property not
immediately available, and the improvement of which, under existing circumstances involves heavy outlay, when superior facilities
and in an equally convenient location are easily within their
reach.   The natural advantages of Baltimore as the entrepot for
all traffic to and from the South and West, have been shown in
previous pages, and these advantages must be utilized; but they
can only be thus utilized in connection with the water front which
is rendered available by nature;—the *only* water front now unoccupied, having a depth of water for the largest vessels which
are now employed in the European and South American trade;—
the *only* water front, from which an easy and economical rail line
can be established in connection with the railroads now in operation to the West and South.   The reader will recognize the force
of these remarks by reference to the accompanying map, and he

will see how convenient Curtis' Bay will be as a port, not merely
for the Baltimore and Ohio, but for the Baltimore and Potomac,
Baltimore and Drum Point, Northern Central and Western Maryland Railroads. As this pamphlet, with its accompanying maps,
will doubtless fall into the hands of some who have not an opportunity of making personal investigation into the correctness of
statements furnished about geographical location, &c., it will not
be inappropriate to state in this connection that "The Patapsco
Land Company of Baltimore City" have used every effort to attain
accuracy in details and in statistics; nothing has been intention-
ally exaggerated; due merit has been conceded to other honest
endeavors made in furtherance of the commercial interests of Baltimore, and the sole aim has been to show that their water privileges, if the *last* to be developed, are not the *least* meritorious;
and that the advantages destined to accrue to the city of Baltimore
from the present movement towards providing additional harbor
facilities, are almost incalculable. The following additional statistics relative to Curtis' Bay may be instructive:
Immediately within the mouth of the bay there are 20, 21, 22,
23 and 24 feet of water. At the east point of what is known as
Cabin Branch, there are 24 feet; and at the opposite point close
to the shore there are 21 feet. At the mouth of Curtis' Bay, that
is from Fishing Point to Leading Point, it is one mile in width,
with water varying from 22 to 24 feet. From Ferry Point, mouth
of Cabin Branch to Sledd's Point, on the opposite side of the bay,
it narrows to 2,000 feet, but immediately expands to more than
half a mile wide, being nearly twice the width of the Patapsco at
the Lazaretto, *inside* of which is the harbor of Baltimore, the property of the Canton company and Locust Point. Vessels drawing
19 feet of water *cannot* enter the harbor of Baltimore, the depth
at the Port Warden's Line being but from 8 to 18 feet. This last
fact is pertinent and significant.
Again, it is claimed by "The Patapsco Land Company," that the
construction of the Maryland and Delaware Ship Canal will have
a marked influence on the success of their harbor at Curtis' Bay,
and in this respect they are right; because the great length of the
bay intervening between the ocean and the shipping wharves is a
great annoyance and cause of delay to sailing vessels in which, at

any rate for some time to come, a large proportion of the coastwise coal traffic will be undoubtedly carried. Once let this proposed canal be built, shortening the distance of navigation to European and Eastern ports 225 miles, and the harbor of Curtis' Bay will swarm with vessels, and Baltimore, instead of lagging behind in the race for commercial supremacy, will occupy a foremost place; and the geographical short rail line advantages, in connection with low port charges, will more than compensate for the prestige claimed by New York in consequence of being the great monied centre of the North American continent.

It may be fairly presumed that the superior location of Curtis' Bay as a port will be candidly admitted by all who examine the map, or who have fortunately an opportunity to investigate its situation personally; but in developing their property "The Patapsco Land Company" believe that a large and influential city will spring up there; that on the various creeks and indentations of the bay manufactories of every description will be established, and that on the rolling land rising gradually back from the bay, which commands magnificent views of the Chesapeake Bay, and the lands on both sides of the river, country residences will be erected; in fact, the proposed town of Pennington may become for the city of Baltimore what New Brighton, Hoboken and Brooklyn are for New York. Among the arguments adduced in favor of the position that Baltimore was destined to become a large manufacturing centre, was the existence at that point of *cheap* fuel, *cheap* rents and a *cheap* market. The same argument is applicable to Curtis' Bay. To this point will converge the coal of the Cumberland Region and that of the Lykens Valley, here contiguous to the water and with every facility for handling economically the raw material or shipping his goods the manufacturer will erect at a cheap ground rent his factory, or the artizan on similar favorable terms his house; here will be the terminus of the Baltimore and Drum Point Railroad and the varied market productions of Western and Southern Maryland;—its fruit, its butter, its eggs, its vegetables and fish, to say nothing of its oysters and game, will find at all times a ready sale. Here it may be noted that the Curtis' Bay property has within itself so to speak a great element of wealth in the valuable clay which is found at various places in the tract and is estimated according to the opinion of experts to cover many hundred acres. Of course the depth of these clay banks can only be ascertained by actual development, but if the banks are only 10 feet thick, they will yield a royalty of $4,500 per acre. More importance may be attached to the clay deposit, from the fact that the supply in the vicinity of Baltimore is rapidly becoming exhausted, and the supply is now not equal to the demand. The Baltimore *pressed* brick commands a very high price in the Southern and Eastern markets and a large trade may soon be built up on the property of "The Patapsco Land Company" the clay having been actually tested and experimented with producing brick of unsurpassed quality and beauty. Again the whole country between Baltimore and Curtis' Bay and further south into Anne-Arundel county is full of iron deposits, found in the clay. The ore from these pockets, when smelted yields metal of the very best quality and justly celebrated for its strength, ductility and tenacity. This iron ore is now carted to Baltimore and finds a ready market, but as soon as this property is developed and a connection by rail established for bringing the coal to tide water, furnaces, rolling mills, foundries, machine shops and car wheel or locomotive manufactories will come into existence;—their owners being mainly influenced by the attractions of *cheap* fuel, *cheap* location and *cheap* material. Saw mills, planing mills and other establishments for the manufacture of furniture will also be erected. Allusion is made to this point because it is believed that the lumber traffic of the Baltimore and Ohio, Western Maryland and Northern Central Railways will ultimately converge to this port, the facilities at Locust Point not being sufficient to accommodate the increase of general traffic. It is stated that large shipments of American walnut are now made to Europe; and as the Chicago line of the Baltimore and Ohio Railroad taps at Auburn a road which traverses between that point and Terre Haute, extensive forests of walnut and other hard wood, there is no reason why shipment in bulk of furniture woods should not become an important item in the traffic from Curtis' Bay to trans-Atlantic ports. Again as has been previously stated on page 33, Oyster, Fruit and Vegetable packing is an important item in the manufacturing industries of Baltimore, and there is no reason why

Curtis' Bay as situated in the immediate vicinity of the great fruit growing and vegetable farms from which the city is at present supplied, and as easier of access to the small vessels employed in the coasting trade should not participate in a manufacture which is remunerative and in which a large number of hands are constantly employed. Lime kilns could also be erected in connection with the oyster canning establishments and the lime could readily be disposed of for agricultural or building purposes, or the shells could be sold to the blast furnaces by which they are used in the smelting of iron. Here the coal oil brought from Parkersburg and Pennsylvania could be refined, tanks erected and suitable accommodations furnished for shipment to Europe ;—vessels with an outward cargo of petroleum would return laden with the marble of Italy or the fruits of the Mediterranean. Here also abattoirs might be erected and packing houses built far away from the other manufacturing establishments in the town at the head of the bay, and here also a large business might be transacted in the manufacture of guano, chemicals and fertilizers. In alluding thus briefly to the reasons why Curtis' Bay is in all probability destined to become equally important as a shipping port and a manufacturing centre, the aim has been to suggest ideas which are in consonance with the experience of other places, which have attained a high stage of development under much more unfavorable auspices and with less facilities. The acquaintance of each individual reader with the idiosyncrasies, so to speak, of his own trade or business will doubtless suggest from the brief resumé of the conveniences at Curtis' Bay where or how he can make a profitable investment, and thus participate in the future prosperity and growth of the new shipping port.

Before concluding this notice of the advantages of Curtis' Bay as a port and manufacturing centre, it may be appropriately noted that the project of making a canal to connect the city of Baltimore with the Chesapeake and Ohio Canal at or near Georgetown, has been resuscitated, and at the last session of the Legislature the "Maryland Canal Company" was reïncorporated. It is believed that the debouchure of this canal will be at Marley's Creek, and Curtis' Bay will naturally be materially benefitted by the volume of traffic which must be coincident with the completion of the

new enterprise. Half a century has nearly elapsed since the
Legislature of the State of Maryland evinced a deep interest in
the construction of a canal which should have Baltimore as a terminus, of a great highway to the coal fields of the Alleghanies,
and passed an act in June, 1825, incorporating the "Maryland
Canal Company," to cut a canal from some convenient point on
the Potomac, intersecting or continuing the Chesapeake and Ohio
Canal to the city of Baltimore. To further the enterprise, a subscription of $200,000 was made at that time by the State. This
subscription was subsequently revoked in 1827, although the feasi-
bility of constructing a canal had been ascertained by a critical
survey instituted under the auspices of the United States Government. Since that time, until the last session of the Legislature
the enterprise has slumbered, although various attempts have been
made to organize different companies to carry through the work of
connecting Chesapeake bay with the Potomac; now, however,
new life has been infused into the project—a company has been
organized with a capital of $1,000,000, and the city of Baltimore
has been authorized to subscribe to, or endorse the first mortgage
bonds of the reïncorporated company to an amount not exceeding
$1,500,000, as to the Mayor and City Council may seem advisable.
As the preamble of the act incorporating the company states that
the route of the proposed new canal is not more than 28½ miles, it
may reasonably be assumed that Marley's Creek is the objective
point of the new work. If the Maryland and Delaware Ship
Canal is constructed, the impetus given to the coal traffic from the
Cumberland region will be immense, and the city of Baltimore
would receive corresponding benefit.

The reader will doubtless be aware that the Chesapeake and
Ohio Canal was originally projected as its title imports to connect
the Ohio River with the Chesapeake Bay, but it has never yet
progressed beyond Cumberland, and it is only within the last six or
seven years that the coal tonnage has assumed a magnitude to
insure a sufficient revenue for defraying the ordinary operating
expenses, much less to meet the interest obligations of the company.
The tonnage has however increased from 482,325 tons in 1868 to
778,802 tons 1873, and although the universal depression in trade
incident to the panic has effected prejudically the earnings of the

company during the current year, still they have the satisfaction
of knowing that their property is now in first class condition and
that the arrearages of interest are being gradually paid off, also
that when the Western Maryland Railroad which connects with
the canal at Williamsport is completed through to tide-water, a
large coal traffic must be sent by that route to Baltimore and be
shipped from there to Eastern ports and to the West Indies and
South America. It may also be noted that steps will shortly be
taken to extend the canal from Cumberland west to the slack water
at the head of the Monongahela River. Surveys have been made
of a greater portion of the route under the auspices of the United
States Government and the route by the North Branch of the
Potomac has been declared practicable, but the appropriation was
not sufficient to make instrumental surveys of what is known as
the Will's Creek route, and hence no definite estimate has been
arrived at of the cost of completing the canal to a connection with
the Western waters, but it is believed that it would in no case exceed $20,000,000.

A report, presented to Congress from the Senate committee on
transportation routes to the seaboard, indicates that the only solution of the cheap transportation problem was to be found in the
establishment of through water lines from the West to the East,
and it was claimed that such water facilities were imperatively
demanded by the producers of the West, and by the mining and
manufacturing interests of the whole country. It is not within
the province of this pamphlet to discuss whether the position
taken by the Senate committee was correct or the reverse; it is
sufficient to know that a strong pressure will be brought to bear
on the general government to induce them to aid these schemes of
inland water transportation, and should their efforts be successful;
should the Chesapeake and Ohio Canal be completed through and
a connection by canal established between the Potomac and the
Patapsco, near Baltimore, Curtis' Bay will assume additional importance as a shipping and manufacturing centre, and a new
impetus will be given to the petroleum traffic which can never be
anticipated under other and existing circumstances.

(VIII.) PRESENT PLAN OF THE PATAPSCO LAND COMPANY OF BALTIMORE CITY FOR DEVELOPING THEIR PROPERTY.

It has been previously stated that the Curtis' Bay property had been held for a long term of years by its present owners;—held under a firm and abiding conviction that the intrinsic value of such property would be eventually appreciated, and its necessity in connection with the commercial and manufacturing development of the city of Baltimore candidly recognized. The present stockholders of "The Patapsco Land Company" have organized with a capital amply sufficient for carrying out well conceived plans for harbor facilities and terminal improvements, while the active direction and management are in the hands of those who are socially and commercially identified with the past and present history of the city of Baltimore, and whose names and character are guarantees that *what is done will be well done*, and that in every movement the commercial and manufacturing interests of their native city will be aggrandized. A plan of the property will be found in connection with this pamphlet, and on it are delineated on a small but accurate scale some of the improvements now contemplated, and in the execution of which the management will employ the most able and most experienced talent of the country. Special allusion is made to this fact, as the plans and designs for the docks, piers and wharves emanate from Mr. Simpson, of New York, a veteran expert in works of this character, and one whose talent is attested by the successful construction of docks, elevators, &c., in the sister cities of New York and Philadelphia. Fortunately the location of the property obviates the necessity of any very heavy preliminary expense in erecting a bulkhead all along the water front contiguous to deep water, and the company propose to sink cribs, where the water is twenty-four feet deep, and to fill in from such bulkhead line solid to the main land. This bulkhead will commence at or near what is known as

Stonehouse Cove, and following as straight a line as is compatible with the formation of the shore, will extend to the further end of what is known as " Cabin Branch." Further extensions and additions will be made to this bulkhead as they are demanded by the requirements of an increasing traffic, but in the meantime it is believed that in making this permanent improvement the wants of many years are amply provided for. At right angles with this bulkhead, wharves will be extended to deep water, and will be of such a length as to accommodate vessels of the largest size. Fire proof receiving warehouses will be erected on these wharves, and it has been determined that the uniform space between each wharf shall, as at Locust Point, be at any rate one hundred feet in the clear, thereby securing ample dockage for two vessels lying side by side. Elevators will also be built, and provision will be made for coal wharves, at the most convenient point near the north end of the water front, where the high ground will obviate to a great extent the necessity for extending trestle work far into the bay, from which to dump the coal into vessels. In constructing this bulkhead and wharves, the company propose to treat all the horizontal-side ends and tie timbers above the first foot above water by the American wood carbolising process, and thereby secure a permanent character for all their structures. Experience shows that timbers above the first foot above water not treated in this manner show decay in seven years, and in some instances become so rotten as to be blown away by the winds or torn off by the waves. The managers of " The Patapsco Land Company," profiting by this experience, are determined, even at an increase of original outlay, to prevent, at any rate for a long term of years, any perceptible deterioration of their property. Along the whole length of this bulkhead, and running back three hundred feet from the inside thereof, will be a quay three hundred feet in width, capable of accommodating, if necessary, twenty-one tracks, and from such quay there will be switches or turnouts to every wharf, special care being taken in the construction of such wharf or wharves that room shall be left on each side for a single railroad track. It is proposed that on the ground at the back of this three hundred feet, on the side facing the entry to Curtis' Bay, warehouses should be erected by the merchants adopting Curtis' Bay as their shipping port, while on the "Cabin Branch" side, in the rear of the quay alluded to, manufacturing establishments will be located. A large proportion of the coasting trade will be accommodated at Marley's Creek, where, as the reader will observe by reference to the map; there is a good depth of water, and where, by the simple erection of a bulkhead, wharfage facilities can be furnished, and the cargo discharged promptly into warehouses situated contiguous to the bulkhead. At or near this point will be found a convenient place for ship building, and it is believed that one if not more of our well known iron ship builders will be attracted to a locality where they can procure iron of the very best quality at a much lower figure than at Chester or Philadelphia. In connection with this Marley's Creek, it may be noted that, in all probability it will be the debouchure of the Maryland division of the Chesapeake and Ohio Canal, in which event the traffic in coal and other produce converging to that point would be very large. In connection with the tracks laid on the quay, 300 feet in width, extending along the inside of the bulkhead, will be the "Connecting Road" from Pennington to Relay, a distance of about five and a half miles; and the "Baltimore and Western Maryland Railroad" extending from the present terminus of the Western Maryland Railroad, on the west side of the city, to tide-water at Curtis' Bay. It is proposed that the two railroads just mentioned, as well as the Baltimore and Drum Point Railroad, should come into the city on one road, which, as far as laid through the city, shall be 100 feet in width, capable of accommodating, if necessary, seven tracks, and as the ground admits of such an arrangement, the approach to tide-water will be on a gradual decline, the streets being so laid out that they will cross the railroad above grade and not interfere with the conduct of transportation. In the survey of the railroad from Pennington to Relay care has been taken that the gradients shall not exceed twenty feet to the mile, said gradients to be in every instance in favor of the eastward bound traffic. In building this short connecting railroad "The Patapsco Land Company" contemplate profiting by the experience of English railway managers, and all the cross-ties will be treated by the American carbolising process. The life of a tie will thus be guaranteed for at least twenty-four years, and permanent economy will be secured in maintenance. The superstructure will be of steel, and in its fastenings and other appurtenances, more especially in the spikes, which will be screw, will be adapted to the requirements of a heavy traffic. It has not yet been determined whether the company will own its own motive power or allow the railroad companies availing themselves of its line to haul their own trains to tide-water; but however that matter may be settled one thing is certain, only such a tariff of tolls over the road will be charged as will pay a fair interest on the cost in addition to the charges for maintenance, and a reserve fund to compensate for depreciation. Apropos of this "connecting railroad" to be built under the auspices of "The Patapsco Land Company," and to be used in common, under certain general regulations, by the various railroad companies availing themselves of these superior tide-water facilities, it may be noted that it affords to the manufacturer advantages which cannot be over-estimated, because he is not dependent, as in too many instances, on one line for his transportation, whether of raw material or of the manufactured product; he will not be compelled to take the coal of the Consolidated Coal Company at a high figure, when their cars stand side by side in the yard, with those of the Northern Central and Western Maryland Railroads; cotton from the South can come to him equally as well by the Baltimore and Potomac, as by the Washington Branch Railroad. If the tariff on petroleum from Parkersburg or West Virginia be not in accordance with his views, a satisfactory rate can be obtained from the Northern Central on coal oil from the wells in Pennsylvania. The business man will appreciate fully the advantages of such an independent situation, and the management of this "connecting road," not in the interest of one corporation, but of all, will tend in a great measure to foster manufacturing industries, and to attract capital to the new town of Pennington.

In laying out the city proper, in which as the Company owns 1120 acres in fee and 313 acres by perpetual lease, there are 17,196 city lots, it is the intention to provide at once, in the survey and platting for the grades, sewerage, water mains, gas mains, &c., and it will be so laid out, that throughout the whole scheme there will be a strict uniformity, as far as the necessary requisites for cleanliness, health and comfort are concerned. It is not to be expected that all these city improvements can be made at once, but this much will be done now: the correct, prospective grade of each street will be determined, and in regulating the grades perfect sewerage will be kept constantly in view. Relative to water supply for the city, it is proposed to utilize the Patapsco river, and bring the water from some point near Ellicott's Mills to the rising ground at the back of the city, whence it can be readily distributed and will have sufficient fall to be available in any emergency. There may be natural obstacles in the way of procuring an abundant water supply from the source mentioned, but the geography of the country would indicate that such a course were feasible; if not, other plans will be devised for securing what is of paramount necessity for manufacturers and residents, viz: an inexhaustible supply of pure water. As soon as the survey of the city has been completed and correct plans furnished, the land will be thrown open for sale, or will be leased to manufacturers or merchants contemplating the erection of permanent improvements, at a low ground rent. In the interim, until the property which is the furthest removed from the water front is required for building purposes, it will be rented on short leases to market gardeners, who will appreciate the value of rich, arable land, susceptible of a high degree of cultivation, in proximity to the markets of Baltimore and Pennington. Among the various improvements projected by "The Patapsco Land Company," is the establishment of a steam ferry from Curtis' Bay to such landings in the city as would be necessitated by the requirements of business. This is more necessary, in view of the fact that the Baltimore and Drum Point Railroad will, in all probability, have its terminus at Curtis' Bay, and access to the business portions of the city will be an immediate requirement. Should experience, however, demonstrate the unreliability of such a ferry in all kinds of weather, efforts will be made to secure the construction of a FREE bridge over the Patapsco River, and by the establishment of a street car line between Pennington and the present terminus of the Charles Street Railroad, make a direct connection with South Baltimore and the business portion of the city. In reference to this FREE Bridge, it may be noted, that nothing would tend to enhance the value of farming land in Anne Arundel county more than its construction;

because the tolls charged over the only bridge now existing are virtually prohibitory, being at the rate of six cents per passenger and fifty cents for a wagon and pair of horses. If this monopoly were broken up, and free passage or a minimum rate of toll inaugurated, an unprecedented stimulus would be immediately given to agriculture and to fruit raising in the county south of Curtis' Bay, and numbers of the working class would adopt Pennington as their home, and avoid many of the expenses incident to a residence in the city of Baltimore proper.

In stating thus much about the plans of "The Patapsco Land Company" for developing their property, the reader must carefully bear in mind that experience and time may and in all probability will suggest certain modifications, but still one idea will be kept prominently in the foreground, viz: that *what is done shall be well done,* and in every arrangement the interests of the merchant, manufacturer, shipper and artisan will be carefully guarded, in fact the aim is to create a *cheap* port for the commerce of the country; a *cheap* location for the manufacturer; and a *cheap* home for the mechanic or ordinary laborer, and at the same time in making these general improvements for the benefit of the community at large to guarantee for "The Patapsco Land Company" a fair return for their investment and the risk which they have undertaken.

CONCLUSION.

MUCH more might perhaps have been written on this subject, and the superiority of Baltimore as the great entrepot for the commerce of the West, South-west and North-west might have been more graphically portrayed by one who had more ample opportunities for studying its location and its various surroundings, but the aim has been to adhere strictly to the truth, and not to allow the opinion to be warped by mere chimeras of an imaginative brain. Those of our readers who are acquainted with the city of Baltimore, and who by long residence are thoroughly conversant, as they imagine, with its idiosyncrasies, would do well to pay a personal visit to Curtis' Bay and examine for themselves whether the position taken in this pamphlet in reference to its peculiar adaptation for the great shipping port of the city is not tenable and in accordance with the dictates of a sound judgment. The future of Baltimore rests with her citizens—with those high-minded and public-spirited men, who have consistently predicted for her an era of commercial prosperity, and who, in pursuance of a grand and ennobling idea, have tunnelled mountains and spanned mighty rivers with the view of making the traffic of a continent converge to the city of their birth and the home of their brightest aspirations. Can it be imagined that self-interest or other minor considerations will now divert them from the path which has been so consistently pursued in the past? Can it be otherwise than that they will throw their impartial and undivided influence in furtherance of a scheme which will redound to the credit and perpetuate the commercial prominence of Baltimore?

There are others, however, into whose hands this pamphlet may fall, who have merely known Baltimore historically, and who have never studied carefully the geographical advantages which it possesses as a commercial and manufacturing centre. To these the careful perusal of facts embodied in this pamphlet, may suggest

7

new ideas, and they may be led to examine personally and see whether these things are so, and whether the "Monumental City," either regarded as a home or a place for the investment of capital, is not fully equal, if not superior, to any other city on the Atlantic seaboard. Should these anticipations be realized, and should a fresh and lively interest be awakened in the present and future of Baltimore, "The Patapsco Land Company" will be amply repaid for the labor and expense which has been incurred in bringing Curtis' Bay, and its advantages of location, to the notice of an intelligent and appreciative public.

CERTIFICATE OF INCORPORATION

OF

The Patapsco Land Company

OF BALTIMORE CITY.

THIS CERTIFICATE, made this twenty-fifth day of September, in the year eighteen hundred and seventy-four, by Joseph W. Jenkins, Hiram Kaufman, Joshua Hartshorne, William S. Rayner and William C. Pennington, all being citizens of the United States, and of the State of Maryland,

Whereas the parties aforesaid are desirous of becoming incorporated under the name and for the purposes hereinafter stated, under the provisions of the Maryland Code of Public General Laws, in relation to Corporations.

Now, therefore, they do hereby certify: 1. That the names in full and places of residence of the said Applicants, are as follows :

Joseph Wilcox Jenkins, resides at No. 97 Monument Street, in the City of Baltimore.

Hiram Kaufman, resides at No. 432 Pennsylvania Avenue, in the City of Baltimore.

Joshua Hartshorne, resides at No. 86 Cathedral Street, in the City of Baltimore.

William Solomon Rayner, resides at No. 316 Madison Avenue, in the City of Baltimore.

William C. Pennington, resides at No. 36 West Eager Street, in the City of Baltimore.

2. The name of the said Corporation will be The Patapsco Land Company of Baltimore City, the said Corporation being formed in the said city.

3. The objects and purposes for which such incorporation is sought, are the buying, selling, mortgaging, leasing, improving, disposing of, or otherwise dealing with, land in the State of Maryland, and the procuring, preparing for market, transporting and selling any products of its lands ; also the acquiring, or constructing and maintaining, selling, leasing, or otherwise disposing of any

bridge, pier, wharf, floating or dry dock, or marine railway, elevator or factory; also the constructing, owning and operating a line or lines of telegraph within the said State; also the navigation of the waters of said State by steam, sail or other boats or vessels, and the transportation of goods and passengers therein.

The term of the existence of the said Corporation shall be forty years from the date of this Certificate.

4. The operations of the said Company will be carried on in Anne Arundel County, Baltimore County and the City of Baltimore, and its principal office will be located in the said city.

5. The Capital Stock of said Company will be the amount of Three Million Dollars, in Thirty Thousand Shares of One Hundred Dollars each.

6. The said Capital Stock will consist of Thirty Thousand Shares of One Hundred Dollars each.

7. The affairs of the Company shall be managed by five Directors, who will be the parties to these presents for the first year.

In testimony whereof the parties aforesaid have hereto set their hands and seals on the day and year first above written.

Witness:

GEO. McCAFFRAY.

JOS. W. JENKINS,   [SEAL.]

HIRAM KAUFMAN,   [SEAL.]

JOSHUA HARTSHORNE,   [SEAL.]

WM. S. RAYNER,   [SEAL]

WM. C. PENNINGTON.   [SEAL]

STATE OF MARYLAND, CITY OF BALTIMORE, *to wit:*

I hereby certify, that on this twenty-fifth day of September, A. D. 1874, before the subscriber, a Justice of the Peace of said State, in and for the City aforesaid, personally appeared Joseph W. Jenkins, Joshua Hartshorne, Hiram Kaufman, William S. Rayner, William C. Pennington, and severally acknowledged the foregoing instrument to be their respective act and deed.

GEO. McCAFFRAY, J. P.

www.ingramcontent.com/pod-product-compliance
Lightning Source LLC
Chambersburg PA
CBHW021945190326
41519CB00009B/1147